BLACK HOLES

Previous Works by Clifford A. Pickover

Mit den Augen des Computers

Computers, Chaos, Fractal (in Japanese)

Chaos in Wonderland

Computers and the Imagination

Computers, Pattern, Chaos and Beauty

Fractal Horizons

Future Health

Keys to Infinity

Mazes for the Mind: Computers and the Unexpected

Frontiers in Scientific Visualization (co-authored with S. Tewksbury)

The Pattern Book: Fractals, Art, and Nature

Spiral Symmetry (co-authored with I. Hargittai)

Visions of the Future: Art, Technology, and Science in the 21st Century

Visualizing Biological Information

BLACK HOLES
A Traveler's Guide

Clifford A. Pickover

John Wiley & Sons, Inc.
New York ■ Chichester ■ Brisbane ■ Toronto ■ Singapore

This text is printed on acid-free paper.

Copyright © 1996 by Clifford A. Pickover
Published by John Wiley & Sons, Inc.

Design, composition, editorial services, and production management provided by
Professional Book Center, Denver, Colorado.

Library of Congress Cataloging-in-Publication Data
Pickover, Clifford A.
    Black holes : a traveler's guide / Clifford A. Pickover.
        p. cm.
    Includes bibliographical references.
    ISBN 0-471-12580-6 (cloth : alk. paper)
    1. Black holes (Astronomy). 2. Black holes (Astronomy)—Computer
programs. I. Title.
    QB843.B55P53 1996
    523.8'875—dc20   94-41336

Printed in the United States of America

10 9 8 7 6 5 4 3 2 1

This user's guide is dedicated to all
fellow explorers of black holes,
cosmological doughnuts, and parallel universes.

The engines are humming,
The stars streaking.
The journey begins.

Contents

I became possessed with the keenest curiosity
about the whirl itself. I positively felt a
wish to explore its depths, even at the
sacrifice I was going to make; and
my principal grief was that I
should never be able to
tell my old companions
on shore about the
mysteries I
should

s

e

e

—Edgar Allan Poe
"A Descent into the Maelstrom," 1840

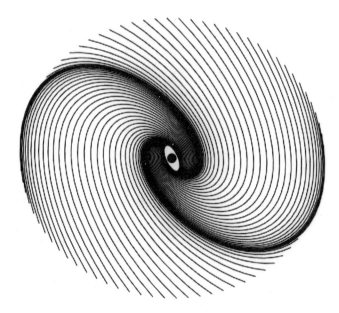

Preface

Siyah-Chal

Siyah-Chal, a black hole horrible beyond imagination, festers in the heart of Tehran, Iran. Known to the rest of the world as the "Black Pit," this nineteenth-century subterranean dungeon is dark, damp, and dismal. It never knows the rays of the sun. Siyah-Chal started its macabre history as a water reservoir for a public bath but later was converted to a prison. Few people who were cast into Siyah-Chal survived. Their screams of terror from the twilight tomb could hardly pierce its massive walls. As if this dark imprisonment were not sufficiently cruel, the chief jailer forced specially selected prisoners to wear one of two dreaded chains around their necks. One chain was known as Qara-Ghuhar. The other chain was called Salisil. Each weighed more than 100 pounds.

> No ray of light was allowed to penetrate
> that pestilential dungeon [Siyah-Chal]
> or to warm its icy-coldness.
> —Baha'u'llah

ix

The dismal history of Siyah-Chal is beyond the ability of most of us to comprehend, yet historians do not doubt its authenticity. To me, Siyah-Chal aptly symbolizes the subject of this book: astronomical black holes that devour everything entering their hungry maws. This sinister feature of black holes has been widely publicized by the popular press; however, black holes have many features, making them much more than cosmic jails that imprison matter and even light. They are also time machines, and may open doors to parallel universes. This book explores these facets of black holes and more.

Abandon Hope, All Ye Who Enter

Since 1960 the universe has taken on a wholly new face. It has become more exciting, more mysterious, more violent, and more extreme as our knowledge concerning it has suddenly expanded. And the most exciting, most mysterious, most violent, and most extreme phenomena of all has the simplest, plainest, calmest, and mildest name—nothing more than a black hole.
— Isaac Asimov, The Collapsing Universe, 1977

Astronomers may not believe in hell, but most believe in ravenous, black regions of space in front of which one would be advised to place a sign, "Abandon hope, all ye who enter here." Dante's caution at the entrance of hell, as astrophysicist Stephen Hawking has suggested, would be an appropriate warning message for travelers approaching a black hole.

Most scientists believe that these cosmological hells truly exist in the centers of some galaxies. These galactic black holes are collapsed objects having millions or even billions of times the mass of our sun crammed into space no larger than our solar system. The gravitational field around such objects is so great that nothing, not even light, can escape their tenacious grip. Anyone who fell into a black hole would plunge into a tiny central region of infinite density and zero volume, and the end of time.[1]

As some historical background, black holes come in many shapes and sizes. Just a few weeks after Albert Einstein published his general relativity theory in 1915, German astronomer Karl Schwarzschild made exact calculations of what is now called the Schwarzschild radius. This radius defines a sphere surrounding a body of a particular mass. Within the sphere, gravity is so strong that light, matter, or any kind of signal cannot escape. In other words, anything that approaches closer than the Schwarzschild radius will become invisible and lost forever. For a mass equal to our sun's, the radius is a few kilometers. For a mass equal to the Earth's, the Schwarzschild radius defines a region of space the size of a walnut.

So far I have made black holes appear frightening; however, they are also a source of infinite mystery and even beauty. In fact, if I had to choose the place and manner of my death, I would certainly elect to plunge into a black hole. Black holes are essentially places that are gone from our universe. And within their hidden boundaries, time and space become magically intertwined.

Sometimes I dream of disappearing from our universe close to the speed of light as I fall into the tiny central singularity where all the laws and properties of our physical universe shatter, where gravity turns time and space to subatomic putty, and God divides by zero . . .

Travel through Time and Space

> The healthy side of the black-hole craze is that it reminds us of how little science knows, and how vast is the realm about which science knows nothing.
> — Martin Gardner, "Seven Books on Black Holes," 1981

> There are stars in our universe which one cannot see.
> — Jean Audouze

This book will allow you to travel through time and space, and you needn't be an expert in astronomy or physics. To facilitate your journey, I start most chapters with a dialogue between two quirky explorers who experiment on a black hole in outer space. You are the captain of a ship, a teacher, and a black-hole enthusiast. Your able student is a scolex, a member of a race of creatures with strong diamond bodies. The scolex, Mr. Plex, will do *whatever* you wish. Use him. Send him out near the black hole to conduct experiments. But be careful: If he goes too close, not even his powerful arterial thrusters can counteract the black hole's massive gravitational pull.

Prepare yourself for a strange journey as *Black Holes: A Traveler's Guide* unlocks the doors of your imagination with thought-provoking mysteries, puzzles, and problems on topics ranging from parallel universes to time travel. This book is resource for science-fiction writers, a playground for computer hobbyists, and an adventure and education for beginning physics and astronomy students. Each chapter is a world of paradox and mystery. The various experiments in each chapter are accompanied by short listings of computer code in an appendix. Computer hobbyists may use the code to explore a range of topics: gravitational space curvature, event horizons, tidal forces, gravitational lenses, general relativity, time travel, space warps, blueshifts, gravitational recoil, binary spiralling black holes, quantum foam, wormholes in space-time, and much more. However, the brief computer programs are just icing on the cake. Those of you without computers can still enjoy the journey and conduct a range of thought experiments. Readers of all ages can study black holes with just a calculator.

There are many excellent books on black holes, and these are listed in the "Further Reading" section at the end of this book. So, why another book on black holes? I have found that most black-hole books on the market today have a particular shortcoming. They are either totally descriptive, with no formulas with which readers can experiment—not even simple formulas—or the books are so full of complicated-looking equations that students, computer hobbyists, and educated laypeople are totally turned off. At the latter extreme are books and papers with esoteric mathematical formulas and jargon such as "Riemann curvature tensors," "super-hamiltonians," and "Regge calculus." It's enough to make ordinary readers feel that they are 50 IQ points behind the times. Astrophysicist Stephen Hawking himself mentioned the following old saying of book publishing in his wonderful book *A Brief History of Time*: "Someone told me that each equation I included in the book would halve the sales. In the end, however, I *did* put in one equation, Einstein's $E = mc^2$. I hope this will not scare off half my potential readers." In *Black Holes: A Traveler's Guide*, most chapters focus on a single equation and give sufficient information so readers can implement ideas with a hand calculator. My current book, therefore, permits general readers to explore and understand black holes and parallel universes.

Imagery is at the heart of much of the work described in this book. To understand what is around us, we need eyes to see it. Computers with graphics capability can be used to produce visual representations from myriad perspectives. In the same spirit as my previous books, *Black Holes: A Traveler's Guide* combines old and new ideas—with emphasis on the fun that the creative person finds in doing, rather than in reading about the doing.

The Black-Hole Smorgasbord

A black hole is nothing, and if it is black, we can't even see it. Ought we to get excited over an invisible nothing?
— Isaac Asimov

As in all my previous books, you are encouraged to pick and choose from the smorgasbord of topics. Many chapters are brief and give you just a flavor of an application or method. Often, additional information can be found in the referenced publications. To encourage your involvement, computational hints and recipes for producing some of the computer-drawn figures are provided. For many of you, seeing program code will clarify concepts in ways mere words cannot.

I have created all the computer graphics images in *Black Holes: A Traveler's Guide* and have provided a brief description of the color plates toward the center of the book. Some information is repeated so each chapter contains sufficient background information, but I suggest you read the chapters in order as you and Mr. Plex gradually build your knowledge. As I alluded to in the last section, the basic philosophy of this book is that creative thinking is learned by experimenting. In fact, many of the exercises are of the "stop-and-think" variety and can be explored without using a computer.

I considered several book titles before settling on the current one. Initially, I chose *Hawking for Hackers*, but I was not sure Dr. Hawking would appreciate this one. I almost titled the book *There Are Holes in Our Universe*, to emphasize recent evidence that black holes are not just theoretical ideas in the minds of astronomers, but really are out there in the universe. In fact, black holes are probably commonplace. In the summer of 1994, the Hubble space telescope provided some of the most impressive evidence to date that black holes are real. The evidence consists of a rotating gas disk at the heart of a large galaxy. The gas is orbiting a mass about two-and-a-half billion times the mass of our sun, and this mass is almost certainly a black hole. (See chapter 1 for more on this behemoth hole in the universe.)

In closing, let me remind readers that humans are a moment in astronomic time, transient guests of the Earth. Maybe our minds have not sufficiently evolved to comprehend the mysteries of space and time embodied by black holes. Perhaps our brains, which evolved to make us run from lions on the African savanna, are not constructed to penetrate the mathematical mysteries of collapsed stars. And only a fool would try to compress decades' worth of astrophysical knowledge, from Einstein to Hawking, in just a few pages of short experiments. We proceed . . .

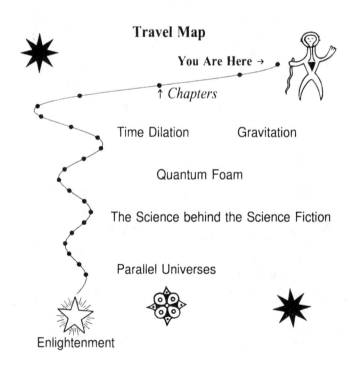

Travel Map

You Are Here →

↑ *Chapters*

Time Dilation   Gravitation

Quantum Foam

The Science behind the Science Fiction

Parallel Universes

Enlightenment

Acknowledgments

I owe a special debt of gratitude to Dr. Kip Thorne, cosmologist and black-hole guru, for his wonderful past books and papers from which I have drawn many facts about black holes. In my book, I use a "Socratic" exchange, or elenchus, between two explorers of a black hole. (The term *elenchus* refers to the method pursued by Socrates of eliciting truth by means of short questions and answers.) Dr. Thorne also used a Socratic approach in one chapter of his book *Black Holes and Time Warps*.

I thank Clay Fried, Arlin Anderson, Robert Strong, Warren Anderson, Marc Hairston, William Lawson, Tom Jackman, and Joseph Pycior for useful comments.

I created all the figures in this book using my own computer programs, many of which are given in the appendix titled, "Smorgasbord for Computer Junkies." To render the data, I wrote several graphics programs and also used IBM Visualization Data Explorer, VossPlot, Galaxy, and GalSoft. Some of the images were computed using an IBM Power Visualization System, a parallel graphics supercomputer. Other images were created using an IBM RISC System/6000. Despite the fact that I used such powerful systems, you should be able to generate many of the images using other software running on personal computers.

C H A P T E R 1

---

*Astrophysicists have the formidable privilege of having the largest
view of the Universe; particle detectors and large telescopes are
today used to study distant stars, and throughout space and time,
from the infinitely large to the infinitely small, the Universe
never ceases to surprise us by revealing its structures little by little.*

— Jean-Pierre Luminet, *Black Holes*

*Almost anything some astronomers suggest about a black hole is
denied by other astronomers.*

— Isaac Asimov

How to Calculate a Black Hole's Mass

The year is 2090 and you are chief curator of an intergalactic zoo. Nicknamed *Theano*, your large ship carries animals from several star systems. Currently on your viewscreen is a black hole with streams of interstellar gas being pulled into its dark interior. X rays pour out from the vicinity of the hole, caused by the rapidly vibrating, in-falling atoms of gas.

You turn to your first officer. "Mr. Plex, I want you to help determine the mass of this black hole."

Your first officer is a scolex, a race of creatures with diamond-reinforced exoskeletons that allow them to explore outer space with little consequence to their health.

"Yes, sir," Mr. Plex says. His slight hesitation is betrayed by the quivering of his forelimbs.

"You are to propel yourself toward the hole and then begin to orbit it."

"Sir, how will this help?" There is a vague metallic twang to his voice.

Your voice is firm. "I will explain when you come back."

Using your powerful telescope, you watch the scolex descend toward the hole. Mr. Plex uses his forelimbs in a kind of sign language to report the amount of time required for one circular orbit around the hole. He also reports the length of the orbit.

After some time, the scolex returns to the ship. His breathing is rapid and heavy as he makes a bassoonlike sound.

You wait a few moments for him to catch his breath. "Mr. Plex, the formula we'll use to approximate the mass of the black hole is derived from Newton's law of gravitation and

3

from the fact that the speed of a circularly orbiting object is the circumference divided by the period.[1] We have:

$$M_b = \frac{C_0{}^3}{2\pi G P_0{}^2}$$

where M_b is the mass of the black hole. The variable C_0 is the circumference of your circular orbit around the hole. P_0 is the period of the orbit, that is, the time required to circle the hole once. G is Newton's gravitational constant 1.327×10^{11} kilometers[3] per second[2] per solar mass."

The scolex's breathing rate has almost returned to normal. "Sir, I completed one orbit around the hole in 10 minutes. My orbital circumference was 4,500,000 kilometers—approximately equal to the circumference of your sun with a diameter of 4.56×10^9 feet." Mr. Plex pauses.

You nod. "Therefore, the mass of the hole is 303 solar masses, or 303 times the mass of the sun. If the orbital circumference were less, say, 1,100,000 kilometers, then the mass would be 4.4 solar masses. If the period doesn't change, you can see from the formula that as the circumference decreases, so does the mass."

"Sir, may we write a computer program to better understand the relationship between the mass of the hole and the orbital circumference and period?" Without waiting for your answer, the scolex writes a short program and feeds in some values. The figure on the next page appears on your viewscreen.

"Very good, Mr. Plex. The plot shows the mass of a black hole as a function of the period of an orbiting body. Each curved line on the plot corresponds to a calculation performed for a different value of the circumference. The labels 1, 3, 5, and so on indicate the value of the circumference times 10^5. For example, the line labelled with a 3 corresponds to 3×10^5 kilometers."[2]

"Sir, I have one final question for you." Mr. Plex pauses. "I have seen elephants in the intergalactic zoo. How long would it take me to circle a black hole with the mass of an elephant,

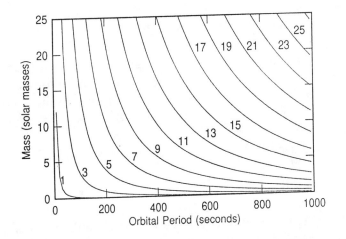

Mass of a black hole as a function of the orbital period. Each curve represents a different orbital circumference value (see text).

given the circumference value I reported of 4,500,000 kilometers? The mass of an elephant (4,000 kilograms) is quite small when compared to the sun's mass of 1.98×10^{30} kilograms."

You smile, tap a few buttons near the viewscreen, and display an elephant grazing in the African wing of the ship. "Mr. Plex, shall we launch the elephant into space and give it a try?"

What is the answer to Mr. Plex's question?

THE SCIENCE BEHIND THE SCIENCE FICTION

There Are Holes in Our Universe

The black hole you and Mr. Plex have found is almost certainly one of many that exist in the universe. Today, there is significant evidence that black holes are not simply interesting theoretical constructs; however, because black holes are not readily observable, astronomers can only collect data that infers the existence of black holes. For example, Cygnus X-1, an intense X ray source in the constellation

Cygnus, is almost certainly a black hole. The X-ray emission is the prime clue. Matter torn off from a neighbor star forms a swirling disk in which gravitational energy is turned into heat and produces X rays. By observing the orbit of the visible neighbor star around the "invisible" black hole, scientists conclude that the hole must be at least three times as massive as the sun and packed into a sphere no more than about 15 million kilometers across. If we were to reduce the companion star to the size of a football, Cygnus X-1 would be the size of a grain of sand orbiting a few centimeters above the football's surface.

More recently, a black hole was identified in the core of the galaxy M87. In 1994, scientists used the Hubble space telescope to study M87's inner regions and discovered a previously unknown disk of gas whirling at a speed of 750 kilometers per second (1.2 million mph), about 25 times the velocity at which the Earth orbits the sun. From the rapid motion of this cosmic whirlpool, scientists estimate that the gas is orbiting a central mass possessing around 2.5 billion solar masses. Moreover, the disk is oriented roughly perpendicular to gas jets that shoot from the center of the galaxy—which is what astronomical theory would predict for the behavior of energy around a rotating black hole. Scientists estimate that the hole measures about 5 billion kilometers in radius. This small size and rapidly spinning surrounding disk rule out most astronomical objects except for a black hole.

Interestingly, today many astronomers believe that black holes inhabit the center of most large galaxies. Astronomers at the Johns Hopkins University in Baltimore recently found strong evidence for another black hole at the center of the spiral galaxy Andromeda. University of Arizona astronomers, using a new high-speed, infrared camera mounted on a Kitt Peak telescope, believe they have discovered a black hole near the center of our own galaxy, the Milky Way.

The most exciting evidence for black holes comes from ten radio telescopes (collectively known as the Very Long Baseline Array, or VLBA). The VLBA allowed scientists in 1995 to peer into the spiral galaxy NGC 4258. Researchers measured the swirling motion of a gas disk orbiting the galactic core. From this motion, clocked at 900 kilometers per second, astronomers concluded the core has a density greater than 100 million suns per cubic light year—exceeding the density of any galactic center ever measured. This is one of the strongest cases yet for a supermassive black hole in a galactic nucleus.

How Many Black Holes Exist?

How many black holes inhabit our galaxy? No one is sure. Dr. Robert Wald has suggested that supernovas (exploding, dying stars) occur in our galaxy at the rate of several per century. *If* a high fraction of them result in the formation of a black hole, there could be as many as 100 million black holes in our galaxy. (The evidence for the existence of *any* black holes requires continuous diligent research and observation, but appears to be growing ever stronger from year to year as shown in the previous section.)

Stephen Hawking believes that, in the long history of the universe, many stars must have burned all their nuclear fuel and collapsed. The number of black holes, according to Hawking, may be greater than the number of visible stars, which totals about a hundred thousand million in our galaxy alone.

Black Holes' Crushed Cousins

Some of you may be interested in black holes' crushed cousins: white dwarfs and neutron stars. Stars are born when a large amount of hydrogen gas starts to collapse in on itself due to gravitational attraction. As the star coalesces, it heats up and produces light, and helium is formed. Eventually the star runs out of hydrogen fuel, and it starts to cool off and contract somewhat. For relatively small stars, the Pauli exclusion principle keeps the electrons in a star sufficiently separated to prevent the star from contracting further after it has spent its fuel. In other words, the electrons counteract the crushing gravitational force. However, for stars more than about 1.5 times the mass of the sun (a mass known as the Chandrasekhar limit), this repulsive force would not be enough to stop stellar collapse.

Let me give some examples. If a star is about one solar mass or less, electron repulsion halts complete gravitational collapse, resulting in a stable graveyard state called a white dwarf. A white dwarf is maintained by the exclusion-principle repulsion between the electrons in matter. White dwarfs have radii of a few thousand miles, and we can observe a large number of white dwarf stars in the heavens. One of the first to be discovered is a star orbiting around Sirius, the brightest star in the night sky.

Aside from white dwarfs, there is another possible graveyard state for stars. If a star is about 1.5 times more massive than the sun, the resulting object formed would be a neutron star. A neutron star is maintained by the exclusion-principle repulsion between neutrons and protons, rather than between electrons. Neutron stars have radii of only around ten miles.[3]

Around 1928, Indian graduate student S. Chandrasekhar was able to show that the exclusion principle could not stop the collapse of a star more massive than about 1.5 times the mass of the sun. Stars with greater masses, in some cases, may explode and manage to get rid of enough matter to reduce their masses below the limit and avoid gravitational collapse. Other large stars may form black holes.

Therefore, white dwarfs maintain the strongest resistance against gravity; black holes, the weakest; and neutron stars, somewhere in between. Interestingly, neutron stars are formed of an extraordinary material called neutronium. Neutronium is so dense that a chunk the size of a thimble would weigh about 100 million tons. The sun, if squashed into a neutron star, would be only a few miles across; the Earth, just a few inches. The density of neutronium is 10^{14} times greater than ordinary solid matter.

Who Coined the Term *Black Hole*?

I have found that few people realize that the term *black hole* is of very recent origin. It was first informally applied to collapsed stars by John Wheeler in 1967, in a meeting of the American Association for the Advancement of Science. The term *black hole* first appeared in print in the January 1968 issue of *American Scientist*. Prior to 1968, the terms *frozen star* and *collapsar* were often used.

The basic idea for black holes occurred first to the English clergy-man John Michell as early as 1783. He started one of his publications with, "Let us now suppose the particles of light to be attracted in the same manner as all other bodies which we are acquainted." In this paper he calculated the characteristics of a massive body with the same density as the sun but with a gravitational pull of sufficient strength to ensure that "all light emitted from such a body would be made to return towards it, by its own proper gravity." Michell suggested that there might be a large number of stars in our universe that gravitationally trap light.

C H A P T E R 2

But little time will be left to me to ponder upon my destiny! The circles rapidly grow small—we are plunging madly within the grasp of the whirlpool—and amid a roaring and bellowing and thundering of ocean and of tempest, the ship is quivering—oh! God!—and . . . is going down!

— Edgar Allan Poe, "Manuscript found in a bottle"

The brain is a three pound mass you can hold in your hand that can conceive of a universe a hundred-billion light-years across.

— Marian Diamond

The Black Hole's Event Horizon Circumference

You are sitting upon the bridge of the intergalactic zoo, sipping a cup of steaming coffee and gazing out at the black hole on the viewscreen. Mr. Plex, your scolex and your first officer, suddenly turns to you and starts to scream. "Sir, have you lost your mind?"

He is screaming because you have asked him to leave the spaceship and begin orbiting close to the black hole in a circular orbit 4,000 miles (6,437 kilometers) in circumference. Even though his diamond exoskeleton protects him from the hazards of outer space, he does not wish to go so close to the hole that he will be gravitationally sucked in with no possibility of escape.

"Mr. Plex, you have nothing to worry about. I am not asking you to go within the event horizon of the hole, from which nothing, not even light, may escape the huge gravitational pull. This black hole's horizon circumference is certainly smaller than 4,000 miles."

Mr. Plex takes a deep, raspy breath. "Sir, how can you know this?" Perhaps there is awe in the scolex's eyes as he takes a sip of his diamond-laced beverage.

You gently place your coffee cup on a nearby table. "I know this because the formula for the horizon circumference is:

$$C_h = \frac{4\pi G M_h}{c^2}$$

Here, G is Newton's gravitational constant, 1.327×10^{11} kilometers^3 per second^2 per solar mass and c is the speed of light, 2.998×10^5 kilometers per second."

The scolex nods. "Sir, just to be on the safe side, I'll write a computer program to determine the circumference of the event horizon. From our last expedition we know that the black hole has a mass 303 times the mass of your sun."

Mr. Plex programs rather slowly because his forelimbs are not designed for human keyboards. Nevertheless, after much effort, Mr. Plex prints the following output showing the event horizon circumference for a few selected values of the hole mass:

```
Hole Mass: 1   solar mass   Circumference: 18.55313  kilometers
Hole Mass: 101 solar mass   Circumference: 1873.866  kilometers
Hole Mass: 201 solar mass   Circumference: 3729.177  kilometers
Hole Mass: 301 solar mass   Circumference: 5584.488  kilometers
Hole Mass: 401 solar mass   Circumference: 7439.801  kilometers
```

For a value of 303 solar masses, the circumference length is around 5,621 km, about the distance from New York City to Los Angeles. This means that a mass 303 times the sun's mass is crammed into a region of space smaller than the United States.

Mr. Plex turns to you and breathes a sigh of relief. "Sir, you are quite right. I can establish an orbit at the circumference you suggested, because I will be outside the event horizon."

"Mr. Plex, did you ever really doubt me?"

The scolex takes a step back. "I suppose we can calculate the radius of the event horizon by dividing by 2π, because the circumference of a circle is $2\pi r$. Right?"

You shake your head in disbelief. "Damn you, Mr. Plex! Haven't I told you that space is so warped near a black hole that this formula cannot be used?" You take a breath. "In fact, the event horizon radius is much larger than we'd compute from $C = 2\pi r$."

The scolex seems confused by your comment.

"Mr. Plex, think of dropping a lead weight onto the center of a circle drawn on a rubber sheet. The weight will stretch the radius of the circle while leaving the circumference unchanged. It turns out we simply can't compute the exact radius of the black hole's horizon." You pause for a second. "Here's another example. Push your finger—claw in your case—into a rubber sheet at the center of a circle. You can push harder and harder, thereby stretching and increasing the radius more and more, without changing the circumference very much at all."

Mr. Plex begins to pace back and forth. "This seems to mean that we can *never* use $C = 2\pi r$, because space is warped by *any* gravitational field."

"You make a good point. However, the distortion is minor in our daily lives. On the other hand, the space inside a black hole, near the singularity at the center, is quite chaotic and extremely warped." You pause. "Even though the twisted region might be bound by a circumference the size of the ring on my little finger, the radius could be millions of miles!"

The scolex's forelimbs begin to quiver. "Sir, perhaps we may think of the space around a black hole as distorting the value of π. I always though of π as a constant, 3.1415..., but maybe the black hole can be thought of as shrinking its value."

Your right eyebrow raises. "I'll have to think about that one." You hear a faint ticking in your head, a virtual time bomb: 3.1415, 3.1414, 3.1413, 3.1412 . . .

The scolex grins. "There's one thing I'm wondering about. If we were to find a smaller black hole with the mass of your sun, it should have an event horizon circumference of 18.6 kilometers (11 miles). This distance is just a little more than the height of Mount Everest. What is the mass of a black hole whose event horizon is a mere 6 feet (0.0018km) in length?"

THE SCIENCE BEHIND THE SCIENCE FICTION

Schwarzschild Radius

You and Mr. Plex were wise to speak in terms of circumferences rather than radii. The geometry of space in the vicinity of a collapsing body will be significantly curved; nevertheless, it is still *convenient* to attribute a value to the radius r of a sphere in curved space, even though r is no longer necessarily equal to the distance of the surface of the sphere to its center. This Schwarzschild radius, r, can be computed from $2GM/c^2$, where G is Newton's gravitational constant, M is the mass of the black hole, and c is the speed of light. It gives the critical size of a star, below which the escape velocity from its surface is the velocity of light. You'll find in many textbooks that the formula for r is condensed to the enigmatic $r = 2M$. Authors often choose the units of mass, length, and time so G and c both have the value 1. Other technical books set only $c = 1$, and they give the Schwarzschild formula as $R = 2GM$.

Recipe for Black-Hole Creation

If you wish to build a black hole, it is necessary to cram a particular amount of mass into a critical volume whose size is given by the Schwarzschild radius. The recipe below should be of some help.

Schwarzschild Radius for Different Objects

Object	Mass	Schwarzschild Radius
Atom	10^{-26} kg	10^{-51} cm
You	100 kg	10^{-23} cm
Earth	10^{25} kg	1 cm
Sun	10^{30} kg	1 km
Galaxy	10^{11} sm	10^{-2} ly
Universe (if closed)	10^{23} sm	10^{10} ly

Note: sm = solar masses; ly = light year

Fact File

- A 10 solar masses black hole has an area of 5,650 square kilometers, roughly the size of a county.
- A 10 solar masses black hole has a surface gravity at its event horizon of 150 billion times that of Earth.

C H A P T E R 3

There are two ways of spreading light: to be the candle or the mirror that reflects it.

— Edith Wharton

In the deathless boredom of the sidereal calm, we cry with regret for a lost Sun . . .

— Jean de La Ville de Mirmont, *L'Horizon chimerique*

Black-Hole Tidal Forces

"*Mon Dieu!*" the scolex screams at you as he gazes at the black hole on the ship's viewscreen. "My body simply won't withstand those kinds of tidal forces."

You feel like slapping Mr. Plex across his diamond face but resist the temptation. "Mr. Plex, get a hold of yourself. Your diamond exoskeleton will protect you from the gravitational forces of the hole. Diamond is the hardest known substance."

"But, sir, my wife—"

"Listen to me, Mr. Plex. We've been through this before, and you are uniquely qualified for this mission." You take a deep breath. "I admit I was wrong to have asked you yesterday to orbit in a 4,000-mile (6,760-kilometer) circumference around the hole. Even though you were outside of the event horizon, and therefore wouldn't have been trapped by the gravitational pull, the difference in the gravitational pull between your head and feet would probably have torn you apart like a piece of Turkish taffy."

"'Taffy,' sir?"

"Today, I want you to propel yourself toward the hole and then begin to orbit it at a safe distance. You'll enter a circular orbit with a circumference of 100,000 kilometers (3.2×10^8 feet). That's about three times the circumference of the Earth. You'll be quite far away from the black hole."

The scolex's forelimbs begin to quiver. "W-will the forces upon me be great?"

For an instant, you feel like telling the scolex, "No greater than your fat ruby wife," but instead you say, "This is easy for us to determine. The force on your body, expressed as the relative acceleration between two extremes of your body—for example, your head and abdomen tip—is:

$$\Delta a = \frac{16\pi^3 GLM_b}{C^3}$$

Here, G is Newton's gravitational constant, 1.327×10^{11} kilometers^3 per second^2 per solar mass, M_b is the black hole's mass, C is the circumference of your orbit, and L is the distance from your head to the end of your abdomen." You take a step back. "This means a smaller creature, such as an ant, would be much less affected by the tidal forces than you or I, if it were to crawl near a black hole."

"Little chance of that, sir."

You nod.

"Sir, why do they call them tidal forces?"

"I suppose it reminds scientists of the gravitational effects that cause tides in oceans."

The scolex nods. "Sir, let me change the subject. We already determined that the black hole on the viewscreen is 303 solar masses."

"Correct, Mr. Plex. That means if you were to orbit in a 4,000-mile (6,760-kilometer) circumference around the hole, as I asked yesterday, the difference in acceleration between your head and your rear end would be a walloping 11,800 g's. To do this computation, I assumed you to be 6 feet tall (0.0018 km) and made use of the fact that 1 Earth gravity (or 1 g) is 9.81 meters per second^2.

"Sir, let me check this out using a computer program."

You watch as the scolex takes out his notebook computer and shuffles in the direction of the ship's tropical arboretum. His multiple legs move in an oddball synchrony, giving him the appearance of a drunken spider. An hour later the scolex comes back with a tattered computer printout clasped in his forelimbs. On it are gravitational force differences for a few lengths:

```
Length:  .0018 km, Tidal Force:  3.6 g
Length:  .0028 km, Tidal Force:  5.6 g
Length:  .0038 km, Tidal Force:  7.7 g
```

A broad grin sparkles upon his diamond lips. "Sir, this is very good. I'll only experience around a 3.6 g difference between my head and abdomen if I orbit at 100,000 kilometers. Certainly I can withstand that."

"Piece of cake," you agree. "After all, in 1960, astronaut Alan Shepard experienced a 12 g force during the re-entry of the Mercury spacecraft *Freedom 7*. And he was a mere human."

"Sir, the 2090 edition of the *Guinness Book of World Records* reports that the highest g value endured on a water-braked rocket sled is 82.6 g for 0.04 seconds by Eli Beeding on May 1958. He was hospitalized for three days."

"Mr. Plex, you never cease to amaze me with your trivia."

The scolex's eyes seemed to glitter. "In 1982, a land diver from the Pentecost Island dived off an 81-foot platform with liana vines attached to his ankles. His body fell more swiftly than 50 feet per second, and the final jerk of the vine transmitted a momentary g force in excess of 110 g."

"Mr. Plex, that's all very interesting, but—"

"Do we have any click beetles in our insect pavilion?"

"Why?"

"Click beetles experience the highest g force of all animals on Earth. They average 400 g when jack-knifing into the air to escape predators. One scientist found a click beetle that survived a peak brain deceleration of 2,300 g." The scolex takes a deep breath. "The beak of a woodpecker hits the bark of a tree with an impact velocity of 13 mph—"

You slap your hands together to silence Mr. Plex. "Why are you telling me all this?"

"To show you that many Earth animals can withstand high g stresses. Perhaps *you* would like to orbit the black hole yourself. Four g's is something your human body could tolerate."

"Very amusing, Mr. Plex. Now please go out and orbit the hole. Your observations will be invaluable."

"Sir, I have something to show you first. I call it an acceler-gram." The scolex pushes a button, and the viewscreen displays a plot with lines shooting outward from a central point. The lines closer to the center are longer than those farther away.

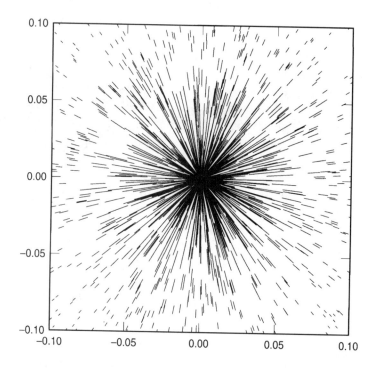

Accelergram. The lengths of the lines correspond to the strengths of the tidal forces acting upon a human positioned at various distances from the black hole. The black hole is positioned at the center of the plot.

You grudgingly admit, "Very pretty."

Light begins to reflect off the scolex's diamond body as it quivers in delight. "This plot schematically shows the relative acceleration between a human's head and feet as he nears a black hole. The bigger the difference in acceleration between your head and feet, the longer is each line segment." Mr. Plex looks you right in the eye. "The longer lines toward the center correspond to your body being pulled apart like a stretched piece of—What did you call it?"

You feel a bit nauseated. "Uh, Turkish taffy."

The scolex smiles. "I have one last question for you. If we were to place the Statue of Liberty (height 1.51×10^2 feet) in the same circumference at which I'll orbit, would the statue be able to withstand the difference in acceleration between its head and feet? What would be the relative acceleration between the top and bottom of your esophagus?"

You roll your eyes. "That's *two* questions, Mr. Plex. I suggest you start out on your mission."

THE SCIENCE BEHIND THE SCIENCE FICTION

How to Explore beneath the Event Horizon

You and Mr. Plex were wise to be careful about the large gravitational tidal forces produced by a black hole. If you were approaching a 10 solar masses black hole with a radius of 30 kilometers, you would be killed long before you reached the horizon, at an altitude of 400 kilometers. However, you could reach the horizon of a 1,000 solar masses black hole, and even be able to explore the *interior* of a 10 million solar masses black hole. The tidal forces at the horizon of this gigantic black hole would be weaker than those produced by Earth, which are already impossible for us to feel. Of course, once you crossed the horizon, you would eventually be pulled into the singularity and then be torn to pieces by the finite tidal forces. This final fate awaits you no matter what the mass of the black hole.

The reason the tidal forces (near the event horizon) are smaller for large black holes than for small holes is as follows. More massive black holes actually have lower density and the exterior space-time is less curved. When the circumference of your orbit is nearly equal to that of the horizon's circumference, we have $C \propto M_h$. Therefore, the formula given in this chapter,

$$\Delta a \propto M_b/C^3,$$

becomes

$$\Delta a \propto 1/M_b{}^2.$$

What If Our Sun Suddenly Turned into a Black Hole?

One of the most interesting questions on gravitation for laypeople involves a macabre scenario where our sun suddenly collapses to a black hole. Although this is not possible from an astronomical standpoint, because the sun does not have sufficient mass to form a black hole (see chapter 1), the *hypothetical* question I like to pose is: What would happen if our sun were to suddenly turn into a black hole containing the same mass as the original sun? Would the Earth, like some gigantic thrown bowling ball, suddenly go crashing into the sun? Would our ocean tides be affected? (Postscript 2 also discusses this topic.)

The answer to both questions is no. The Earth's orbit would be totally unaffected. A sun centered at a particular location in space has the same gravitational effect on the Earth as a black hole of the same mass located at the same position. There would simply be no more light coming to us from the sun if it had collapsed to a black hole. Of course, the effect of this on life would be profound, and we would all perish, but the gravitational effect would not be the cause for our demise.

Beware the Black Tide

Imagine what it would be like to watch the awesome destruction of a star by tidal forces produced by a large black hole. As the star orbits around a black hole, the black hole's gravitational forces act more strongly on the side of the star near the black hole than on the far side. As suggested in this chapter, the tidal force is the difference between these two forces. Let's consider two scenarios. If a star moves in a circular orbit around the black hole, the tidal force remains small, and the star responds merely by adjusting its internal configuration and becoming elongated. In a more destructive scenario, the star moves in a more eccentric orbit, and the tidal forces caused by the black hole increase rapidly as the star approaches the black hole. Eventually, these forces are as large as the forces holding the star together. This happens so quickly that the star cannot adjust its internal configuration, and it deforms catastrophically and is torn apart.

The catastrophic shredding caused by the black tide occurs only when a star approaches the hole within a certain critical radius called the Roche limit. The Roche limit depends mainly on the mass of the black hole. It turns out that a black hole's radius can be larger than the Roche limit. This would be the case for black holes having a mass greater than 100 million solar masses. These obese bullies would only tear apart stars by tidal forces once the stars were *inside* the black hole. We would never see the effects of such gluttony, because all the debris would be sucked into the black hole. Many astronomers today believe that Seyfert galaxies and nearly inactive galactic cores have a central black hole of between 1 million and 100 million solar masses that feed on the remains of stars destroyed by tidal forces. Quasars and the bright nuclei probably harbor a more massive black hole fed by interstellar collisions.

C H A P T E R 4

Some scientists are bothered by infinity, which seems to crop up at embarrassing places in our theories of the universe. A black hole has an infinity at its very center.

— Fred Alan Wolf, *Parallel Universes*, 1988

Science, when it runs up against infinities, seeks to eliminate them, because a proliferation of entities is the enemy of explanation.

— George Zebrowski, *OMNI*, 1994

I like infinities. I believe that infinity is just another name for mother nature. Nature provides infinite possibility all the time.

— Fred Alan Wolf, *Parallel Universes*, 1988

A Black Hole's Gravitational Lens

"It's the animals, sir," the scolex says. "They seem to be very excited today." Mr. Plex pushes an intercom button on the wall and you suddenly hear a warbling sound from the aviary.

You nod and smile. "Perhaps they can sense that we'll be carrying out an unusual experiment today. We want to learn even more about the black hole."

"Unusual?" Mr. Plex says with a nervous warble in his own voice.

"Yes, Mr. Plex. I want you to descend very close to the event horizon today—closer than you've ever been." You pause. "I've determined that your diamond body can indeed withstand the gravitational tidal forces, and your arterial fusion jets should be able to accelerate you back to the ship when your mission is complete."

"*Should* be able to?" If scolexes could sweat, Mr. Plex's body would be glistening.

"Yes, and bring your telescope with you. We can communicate by reading each other's lips."

"Yes, sir."

After an hour passes, Mr. Plex has descended to 1.1 horizon circumferences. You gaze out the ship's portal with your telescope and see him screaming something. You look closer to read his lips.

"Sir, is this close enough? I can't take much more of this." His body is shaking. "The sky above me—it looks very weird."

"Go closer, Mr. Plex," you mouth back to him.

The scolex opens his mouth as if he can't believe you are telling him to go closer to the event horizon. He knows that

25

nothing, not even light, can return to your universe once it has descended below the horizon. The gravitational pull of the black hole would be too great.

"Mr. Plex, don't look down, and everything will be okay."

The scolex's fusion jets begin to fire maddeningly as he tries to slow his fall. There is a wild look in his eyes. "I—I'm at 1.05 horizon circumferences." His forelimbs are being pulled down and stretched like taut guitar strings.

You wave your hand, as though there is nothing to be concerned about. "Mr. Plex, go to 1.01 horizons."

"I cannot do that, sir!"

"Your reward will be great when you return."

Mr. Plex's eyes begin to sparkle with emotion. "The sky— the stars are disappearing! They're turning weird colors."

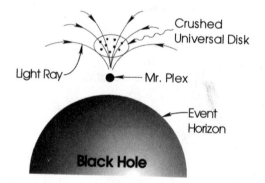

The black hole produces a gravitational lens that squeezes Mr. Plex's view of the universe into a bright circular disk above his head.

You look above Mr. Plex at the stars and galaxies. "They're still here, Mr. Plex. Continue with your mission. Go to 1.01 horizons."

Mr. Plex begins to scream in the madness of extreme agony as he tries to point at the shrinking light from the universe above him. However, the gravitational pull from the hole makes it impossible for him to move his forelimbs.

An hour later Mr. Plex returns to the ship. He is rubbing the sides of his diamond body.

"Sir, you wouldn't believe it. As I descended and looked up and away from the black hole, the starry sky got smaller and smaller, as though I had tunnel vision, as though all the stars were being squeezed into a disk the size of a Frisbee. I felt as though I were trapped in a dark pipe."[1]

"Excellent, Mr. Plex. This confirms my suspicion that the black hole bends the light in such a way as to have made you see all the stars concentrated into a bright, circular spot when you looked up. It should have been like looking up into a bright disk of stars."

"Sir, are you saying the hole acts like a gravitational lens?"

You raise your right eyebrow. "Very perceptive, Mr. Plex. We can even calculate the angular diameter of the circle into which the entire universe is crammed." Your eyes dart around the control room. "Got a calculator? The formula is:

$$a \sim 300\sqrt{1 - C_b/C} \text{ degrees}$$

Here, C_b is the horizon circumference, and C is the circumference at which you are hovering—slightly above the event horizon."

"Sir, what do you mean by 'angular diameter' of the circle? I'd like to draw this on a paper to show my wife."

"That's not too difficult. Let's call the disk of stars overhead the 'crushed universe.' It's really an image of all the stars that orbit the black hole and all the galaxies in the universe. Using a little trigonometry we can convert the angular diameter formula to give the radius of a circle drawn on a piece of paper held about a foot from your eye. This new radius corresponds to the size of the crushed universal circle:

$$r = 12 \times \tan((a/2) \times \pi/180)$$

Here r is the radius in inches, and a is the angular diameter, in degrees, which we get from the previous formula.

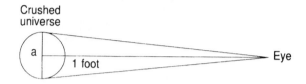

Mr. Plex pulls a notebook computer from a receptacle in the wall and begins to press the computer keys. "Sir, here's the output of a computer program that computes the angular diameter (in degrees) and radius (in inches) for a drawing of the 'crushed universe.' It's a drawing of what you'd see as you looked up while hovering at a circumference 1.01, 1.05, and 1.1 times the event horizon circumference."

Hover Circumference	Angular Diameter	Radius of Crushed Universe
1.01	29.85146	3.198626
1.05	65.46545	7.713259
1.1	90.45355	12.0948

You slap your hands together in amazement. "Mr. Plex, can you imagine that? When you hovered at a circumference 1.01 times the event horizon circumference, the whole universe glowed above you in a star-filled disk having a radius of only 3 inches."

"Sir, I can imagine it. I was there."

"Yes, well, we can write a computer program to draw a schematic representation of the sky as you descended from 1.1, to 1.05, to 1.01 circumferences. You can use the formulas I described and scale a random collection of dots to fit inside the circles. The dots represent stars."

The scolex smiles uncertainly. "Beautiful! We can even make a movie of the universe overhead shrinking as I plunge closer to the hole."

"Quite right. The simulation would be fascinating to watch."

The scolex looks down at his forelimbs. "Uh, sir, you mentioned that I would receive a reward when I returned to the ship."

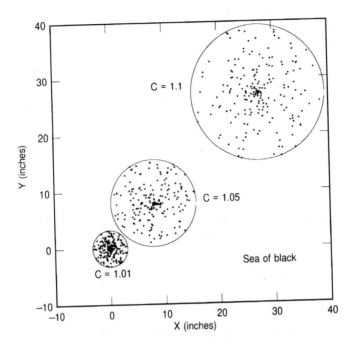

Computer simulation showing the size of the crushed starry disk of the universe as Mr. Plex descends from 1.1 to 1.01 horizon circumferences. Even though he is looking up and away from the hole as he descends, the black hole begins to fill the sky above him. All that is left is a tiny bright circle containing all the stars and galaxies, even those that are located on the other side of the hole.

"Yes. I hooked up a camera to the telescope while you were out studying the black hole." You bring out a nicely framed photograph. It shows Mr. Plex screaming as he descends toward the event horizon. "This will make a nice photo to place on your mantel or by your bed. I'm sure Mrs. Plex will enjoy it."

The scolex remains silent, his facial expression inscrutable. "What would have happened if only my leg pierced the event horizon but the rest of my body was above?"

"Mr. Plex, the laws of physics dictate that nothing below the event horizon can ever return to our universe because the escape velocity needed would have to exceed the speed of light. Therefore, theoretically, your body could have returned, but not your leg. If your leg broke off at the event horizon, you presumably could have come back to the ship. Your leg, however, would plunge close to the speed of light to the singularity at the center of the black hole."

"Tell me more about the singularity. What would it be like to fall into it?"

"At the heart of the black hole is a place where time and space stop, and all the laws of physics break down, a place roughly a hundred billion billion times smaller than an atomic nucleus surrounded by void. But that, Mr. Plex is something we'll consider another day. I suggest you take your wife out to dinner, and tell her all about your interesting day."

THE SCIENCE BEHIND THE SCIENCE FICTION

Boomerang Photons

Gravitational lensing by black holes is often discussed in popular astronomy literature. However, popular books often do not mention that lensing can be so severe as to send photons back to their source, thus producing a gravitational mirror. Beyond Lewis Carroll's wildest looking-glass dreams, black holes create intricate multiple images—if you know where to look. For example, physicist W. Stuckey has written a recent paper on numerical solutions depicting photon trajectories circling a black hole one or two times before terminating at their emission points. This produces a sequence of ring-shaped mirror images. In particular, if you were to approach a nonrotating black hole, you'd actually pass through your mirror image at

$$r = 3M_b G/c^2 = 4.4(M_b/M_s) \text{ km},$$

where M_b is the mass of the hole and M_s is the mass of our sun. For more information, see "For Further Reading." Because gravitational lensing produces a distorted image, some astronomers believe that warped images of constellations may be used to locate isolated black holes.

Fullerene Space Tunnels

The mirror effect described in the previous section produces additional mind-numbing optical effects. For example, if a light ray travels parallel to the surface of a black hole at a distance of exactly 1.5 times the Schwarzschild radius, it will orbit around the black hole in a perfect circle. This means that a space tunnel, circling a black hole at this radius, appears straight to an observer inside the tunnel (Figure 4.1). In actuality, the space structure curves around the spherical hole like a doughnut as seen by a distant observer (Figure 4.2). Because the gravitational field of the hole is so strong, the light you see inside the tube is bent as it circles around the hole. This makes a curved tube actually appear straight.

Astrophysicist Marek Abramowicz has written articles on the force you would feel as you orbited around a space tunnel. No matter how quickly you move, you always feel exactly the same total force pulling inward. For example, if you were motionless, you would feel exactly the same inward force as you would traveling around the space doughnut at the speed of light. There would be no centrifugal force. (See "For Further Reading" for more information.)

Finally, if you hold a lamp within the space tunnel so the lamp rests at the center of the tunnel and then turn around and walk in the opposite direction, you still "see" the lamp in the center. The lamp is never obscured from view by the bend of the tube. As you continue to walk away from the lamp, its image becomes brighter and brighter. In addition, you see multiple images of the lamp as the light from the lamp circulates around the space tunnel many times.

Figure 4.3 shows your view inside the space tunnel when the tube is constructed at a greater distance from the hole, so that light rays travel in nearly straight lines. Here you see the expected curvature of the tunnel.

FIGURE 4.1.   A view from within the circular space ring made of girders and constructed at an altitude of 1.5 times the gravitational radius of the hole. Although the tunnel is curved like a doughnut, it appears straight due to the bending of light rays. The eye in the figure is actually behind you and looking away from you. Because light rays are curved, you can turn away from the eye and still "see" the eye. It is never obscured from view by the bend of the tube. As you continue to walk away from the eye, its image becomes brighter and brighter.

I created the computer graphics of the black-hole space tunnels (Figures 4.1–4.3) using Data Explorer, an IBM graphics software tool. The three-dimensional coordinates of the space tube were supplied

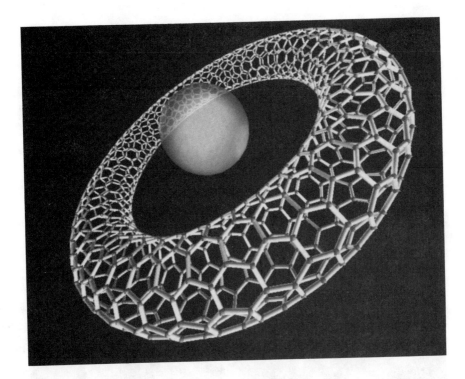

FIGURE 4.2.   An outside view of the space tunnel.

by my colleague Dr. Tom Jackman. The tubes' girder configurations are based on a class of cagelike carbon molecules called fullerenes that are exceptionally strong.

What material should the girders be made from? Arthur C. Clarke's latest note to me indicates he has proposed "space elevators" composed of microscopic tubules. In his 1978 novel *The Fountains of Paradise*, Clarke imagines ultrastrong tubes only a few nanometers wide but hundreds of kilometers long in order to construct an elevator to a tethered space station in synchronous orbit around the Earth. Recently the idea of fabricating tough, microscopically thin fibers from carbon gained credibility with the discovery that sheets of carbon atoms can wrap themselves into microscopic tubes. Richard Smalley of Rice University in Houston notes, "If it were possible to manufacture such tubes in meter lengths, they would be the strongest fibers that we could ever make from anything."

FIGURE 4.3. A view from within the circular space ring constructed at an altitude greater than 1.5 times the gravitational radius of the hole. The curvature of the tunnel is evident because light rays are behaving in a way we are more accustomed to. Far from the hole, light travels in straight lines.

My fullerene space tube surrounding the black hole in this chapter is different from Clarke's idea in that I have arranged huge girders of the space tube to form a gigantic fullerene tube. This arrangement of girders, which mimics the molecular structure, should be remarkably strong. In 1951, Buckminster Fuller designed a free-floating three-dimensional structure called a tensegrity ring-bridge. His dream was to have it installed way out from and around the Earth's equator. Within this "halo" bridge, the Earth would continue its spinning while the circular bridge would revolve at its own rate. He foresaw traffic from Earth vertically ascending to the bridge, then revolving, and finally descending at desired Earth locations.

CHAPTER 5

It didn't end. It went on forever, an infinite landscaped Way constructed from the mathematical fabric of space/time. A hollow singularity that sliced across centuries and galaxies, and into another universe.

— Greg Bear, *Eon*

All we are is light made solid.

— Anonymous

Omnia quia sunt, lumina sunt.
All things that are, are lights.

— Scotus Erigna, 11th century

A Black Hole's Gravitational Blueshift

"The colors!" the scolex shouts over the intercom. "Sir, the stars were the wrong colors!"

You gradually rise from your bed and slap the intercom button. "Mr. Plex, can't this wait until morning?"

"Sir, when I descended closer to the black hole earlier today not only did the stars coalesce into a small region of space, but also all the colors were wrong. When I looked up, away from the black hole, yellow stars started looking green, then blue."

You sigh. "Nothing to worry about. The gravitational force of the black hole caused the color changes. When you hovered at a circumference C above the event horizon with a circumference C_h you saw a decrease in the wavelengths λ of all the light from the external universe:

$$\lambda_{received} = \lambda_{emitted}\sqrt{1 - C_h/C}.$$

This is called a gravitational blueshift. As you looked up, away from the hole, the gravity caused the incoming light to shift to higher energies." You pause. "There's also an opposite effect. If you had shined a flashlight at me from your position near the hole, as the flashlight beam struggled against the hole's gravitational pull it would be gravitationally redshifted to longer wavelengths. I'd see your yellow light turn red."

The scolex is silent for a few seconds. You hear his wife snoring in the background.

"Sir, this means the visible light I saw would have shifted to ultraviolet, and then X rays, as I got very close to the event horizon of the hole. These radiations have shorter wavelengths than visible light."

"Yes, and finally the star light would be shifted to gamma rays, which have a wavelength of 10^{-14} meters. Compare this to yellow light, which has a wavelength of 5.8×10^{-7} meters."

"Sir, can we write a quick computer program?"

"Mr. Plex, it's two in the morning."

You hear Mr. Plex pounding away on a keyboard as his wife asks him to get back to sleep. Her voice sounds like a parakeet being vacuumed out of its cage.

"Sir, my computer program is finished and running. Yellow light is gravitationally shifted to the following wavelengths as one descends closer to the hole and looks up at the light from the stars:

```
Wavelength (meters)
C/Ch            Lambda (emit)    Lambda (rec)
2               5.800000E-07     4.101219E-07   Light
1.5             5.800000E-07     3.348631E-07
1.25            5.800000E-07     2.593839E-07
1.125           5.800000E-07     1.933333E-07
1.0625          5.800000E-07     1.406707E-07
1.03125         5.800000E-07     1.009651E-07
1.015625        5.800000E-07     7.194024E-08   Ultraviolet
1.007812        5.800000E-07     5.106613E-08
1.003906        5.800000E-07     3.617940E-08
1.000008        5.800000E-07     1.602039E-09   X ray
1.000004        5.800000E-07     1.132813E-09
1.000002        5.800000E-07     8.010193E-10
1.000001        5.800000E-07     5.664063E-10
```

"Very good, Mr. Plex. Now let me go back to sleep." Mrs. Plex starts screaming with exponentially increasing frequency and volume. You put your hands over your ears. "Mr. Plex, we'll talk in the morning."

"Sir, wait. There's one thing I've been wondering about. Right at the event horizon your formula gives a wavelength value of zero. What might that mean?"

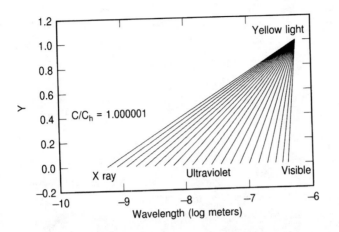

Colorgram showing how yellow light is changed to shorter wave-lengths as Mr. Plex descends closer to the black hole's event horizon. Proceeding from right to left, the ratio C/C_h decreases from 2 to 1.000001.

You punch the intercom button, silencing Mr. Plex and his wife. "Goodnight, Mr. Plex." You put your head on the pillow.

In your dreams you are a bird flying toward the singularity at the center of the black hole. You realize it's a dream, and you wish there were some safe way, in real life, you could sail beyond the event horizon, pierce it, view its singularity where time stops and where the curvature of space-time becomes so severe that you separate from our universe.[1]

C H A P T E R 6

Time is a relationship that we have with the rest of the universe; or more accurately, we are one of the clocks, measuring one kind of time. Animals and aliens may measure it differently. We may even be able to change our way of marking time one day, and open up new realms of experience, in which a day today will be a million years.

— George Zebrowski, *OMNI*, 1994

Eternity is very long, especially near the end.

— Woody Allen

Gravitational Time Dilation

The scolex looks at the watch strapped to his forelimb and then grabs your arm. "Sir, your wristwatch seems to be in error."

You slowly pull apart Mr. Plex's claw and remove it from your arm. "Careful, Mr. Plex, I need this arm."

"But your watch-"

"Mr. Plex, my watch is in perfect working order. Remember that gravity causes time to slow down. Far away from the black hole, where space-time is flat, clocks tick at their normal rates. The other day when you went close to the black hole, I noticed your mouth was moving very slowly. As you moved into regions of increasing gravitational curvature, time slowed for you, and your watch slowed down."

"Why didn't I notice it?"

"Your heartbeat and thinking processes slowed down by exactly the same factor as your wristwatch."

"Does a clock on the ceiling of my home run faster than a clock on the floor? The clock on the floor experiences more gravity."

"Yes, time flows more slowly near the floor."[1]

"*Mon Dieu*! That means the ants creeping on the ground are in a different time flow than humans."

"Mr. Plex, do you need a tranquilizer?"

"It's just-"

You pat Mr. Plex on his back. "The time difference is very small in your examples. But a black hole has a big effect. That's why I didn't let you go too close to the event horizon. I didn't want you returning to the ship only to find me an old man."

"Uh, sir, I thought you were concerned about my safety."

You raise your hand to silence the scolex. "During one second of your time near the hole, millions of years could have flown past our ship. Here's the equation:

$$t_2 = t_1 / \sqrt{(1.0 - C_b/C)}.$$

Here t_2 is the elapsed time when you hover close to the hole, as compared to the elapsed time t_1 far from the hole. An observer near the black hole ages more slowly."

Mr. Plex shoves one of his forelimbs into a cavity of his abdomen and withdraws a notebook computer dripping with oleaginous body fluids. "Found a nice place to store this thing, sir." He looks at the keyboard.

You let out a quick gasp. "Yes, you have."

He begins to type on the keyboard with several of his limbs moving in rapid synchrony. "I've found a more efficient way to interact with the keyboard."

Your eyes are wide. "Yes, I see."

"Sir, my program reports the following:

C/Ch	Time1 (days)	Time2 (days)
2	1	1.414213
1.5	1	1.73205
1.25	1	2.236067
1.125	1	2.999999
1.0625	1	4.123104
1.03125	1	5.744561
1.015625	1	8.062243
1.007812	1	11.35782
1.003906	1	16.0312
1.001953	1	22.6492
1.000977	1	32.01465
1.000488	1	45.26312
1.000122	1	90.50969
1.000031	1	181.0194
1.000008	1	362.0386
1.000002	1	724.0774
1.000001	1	1024

This means that if I hovered at 1.000001 times the event horizon circumference, one day for me would mean 1,024 days for you!"

You grin. "Mr. Plex, you've found the fountain of youth."

"I'm not sure I want to get that close to the black hole."

Your grin widens. "If you do go, just watch that cute little diamond butt of yours. If it were to slip beneath the event horizon while you hovered just millimeters above it . . ."

The scolex takes a step back. "Sir, are you making fun of me?"

"And bring Mrs. Plex with you. You wouldn't want to return only to find her a shriveled up old thing."

"Sir!"

THE SCIENCE BEHIND THE SCIENCE FICTION

The gravitational time dilation you and Mr. Plex experience is quite real. An observer near a black hole ages more slowly than one farther from the black hole. Also, the *proper time* of a clock on the surface of a collapsing star is different from the *apparent time* of the collapse, measured by a distant observer. This is because the surface is accelerating with respect to the distant observer. As J.-P. Luminet and others have noted, the contraction of the star below the Schwarzschild radius happens in finite proper time but in an infinite apparent time. You will never be able to see the formation of a black hole. You will see the collapse get slower and slower as light from the star becomes redshifted and fainter. (See chapter 10 for more information.)

What would happen if Mr. Plex approached too closely to a black hole? Like an ancient ant trapped in amber, Mr. Plex would appear permanently frozen at the event horizon, as his image gradually faded. In actuality, his body would pierce the event horizon as his body plunged toward the singularity.

C H A P T E R 7

Bright star, would I were steadfast as thou art—
Not in lone splendor hung aloft the night,
And watching, with eternal lids apart,
Like nature's patient sleepless Eremite,
The moving waters at their priestlike task
Of pure ablution round earth's human shores,
Or gazing on the new soft fallen mask
Of snow upon the mountains and the moors . . .

    — John Keats (1795–1821),
    "Bright Star, Would I Were Steadfast"

The frontiers of science are always a bizarre mixture of new
truth, reasonable hypotheses, and wild conjecture.

    — Dennis Sutton

Anatomical Dissection of Black Holes

"Mr. Plex, are you very good at dissection?"

"Dissection, sir?"

"You know, surgery. Cutting open frogs with a scalpel."

Mr. Plex takes a step backward. "Sir, on my world, we're all made of diamond. Can't cut a diamond body."

You sigh. "Okay, today you'll get your chance. Follow me." You lead Mr. Plex to a glass cabinet in your cabin, and you unlock the cabinet with a key from your pocket. Inside the cabinet are row upon row of boxes with labels like "static," "charged," and "rotating."

You gesture with your hand to the boxes. "I've had these models ever since I was a kid." You bring out the box marked "static," open it, and reveal something that looks like an edible orange. You carefully place it on your desk. "Today I'll tell you about the structure of three kinds of black holes." You point to the object on your desk. "This is a model of a static black hole."

"Static, sir?"

"It doesn't rotate. This one also doesn't have an electrical charge." You pause, pointing to the skin of the sphere. "This, Mr. Plex, represents the event horizon."

Mr. Plex gingerly touches the smooth surface. "How did it get that name, sir?"

"In a black hole, we have no way of knowing what is happening inside the horizon. We cannot see any interior events. As you know, anything that goes inside this boundary cannot escape the black hole's gravitational pull." You pause. "Some go as far as to say that the region inside the event horizon is dis-

45

connected from our space and time, and it's not part of our universe. Once inside, you can't communicate with anyone on the other side of the event horizon. Not even light can escape."

"Sir, looking at that orange is making me hungry." Flecks of amber saliva appear on Mr. Plex's diamond lips.

"Mr. Plex, control yourself. It's not an orange. It's a model of a static black hole."

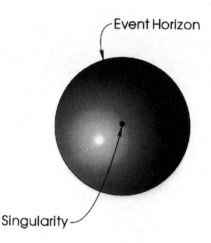

Artistic representation of a static black hole.

"Sir, when I said 'hungry,' I was referring to my hunger for knowledge. What's inside the skin?"

You hand Mr. Plex a scalpel. "Cut it open and take a look."

Mr. Plex grasps the scalpel in one of his claws and begins to pierce the model's event horizon. Is that a shiver of delight you see running along Mr. Plex's forelimbs? Suddenly Mr. Plex takes a step back and yells, "*Mon Dieu!* Sir, it's empty."

"Quite right—except for the tiny speck floating at the center. If this black hole started out as a star, all the mass would be crushed down by gravity into the central speck. In the speck is infinite pressure, density, and space-time curvature. This is where the star is hiding." You pause to increase the drama, and then you whisper, "This is the black hole's singularity."

"But, sir, that means a black hole is nothing but two parts: a singularity surrounded by an event horizon. And there's nothing physical or tangible at the event horizon because all the matter is in the singularity. Is that all there is?"

"Correct. A black hole is empty, except for the tiny singularity and the gravitationally warped space around it." You pause. "Let's move on to the second model." You bring out an object from the box marked "charged."

"Sir, fruits are my favorite food."

You eye Mr. Plex suspiciously. "Here is a model of an electrically charged black hole. Cut it open, Mr. Plex."

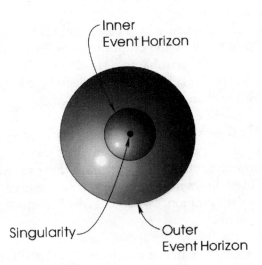

Inner
Event Horizon

Singularity

Outer
Event Horizon

Artistic representation of a charged black hole.

Mr. Plex slices into the event horizon and hits something hard. "Holy—What could this be, sir? The singularity?"

"No, Mr. Plex. You have just cut through the outer event horizon and into the inner horizon."[1]

"An electrically charged black hole has two horizons?"

"Yes. Suppose I started with a static black hole." You point to the discarded rind of the orange model. "If I were to drop electrons into it, it would pick up a charge, and a second event horizon would form and closely surround the singularity. There would now be two places around the singularity where time would appear to stop."

"What happens if we add too much charge?"

"Try it." You hook up a small battery to the model. The inner horizon begins to grow like a balloon as the outer event horizon shrinks. They finally collide. The battery begins to smoke.

Mr. Plex yells, "Sir, it can't take much more of this! What if we continue to try to add more charge?" As though on cue, both event horizons explode and disappear, leaving behind only the tiny floating singularity.[2]

"Now we call the singularity a naked singularity because it is no longer 'protected' by the event horizons." You pause. "Although it's theoretically possible to have electrically charged black holes, it's very unlikely. Nearby oppositely charged particles would neutralize any charge in the black hole."

You look into your cabinet. "Mr. Plex, now for my favorite model." With a flourish, you bring out an object resembling a lemon.[3] "Aside from mass and charge, a black hole can have a third, final property: rotation. Although we don't expect to find charged black holes in the universe, rotating black holes are probably quite common." You hand Mr. Plex the scalpel and he begins to slice.

You gaze at the yellow surface of the model. "The rind on this lemon model is not an event horizon. It's called the static limit. Keep cutting, and I'll explain."

There is a squishy sound. "Sir, I hit something. Is this the event horizon?"

"Yes. You've just pierced the static limit, guided your blade through the ergosphere, and scratched the event horizon." You pause. "If I were to send you out to penetrate a rotating black hole, even before you reached the event horizon it would be impossible for you to stay still, no matter how hard you fired

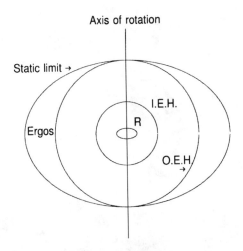

Cross-section of a rotating black hole. (I.E.H. = inner event horizon; O.E.H. = outer event horizon; R = ring singularity)

your arterial thrusters. Although you could use your thrusters to stop yourself from falling into the hole itself, and could even escape back to the space outside the hole, you will be dragged sideways, no matter how hard your rockets blast." You pause again. "Think of yourself in a boat on the side of a whirlpool. The region of space where you are dragged sideways by the rotation of space around the hole is called the ergosphere. It is bounded by the event horizon below and the static limit above."

Mr. Plex's forelimbs quiver. "Outside the static limit, I can remain static with respect to the rotating hole."[4]

"Quite right. It's only inside the event horizon that matter is completely imprisoned."

Mr. Plex continues to cut. "Sir, a rotating black hole is much more complicated than a static one."

You nod. "Obviously."

"I've hit something again."

You peer into the guts of the dissected black hole. "You're now piercing the inner horizon. Just as with the charged back hole, there are again two places around the singularity where

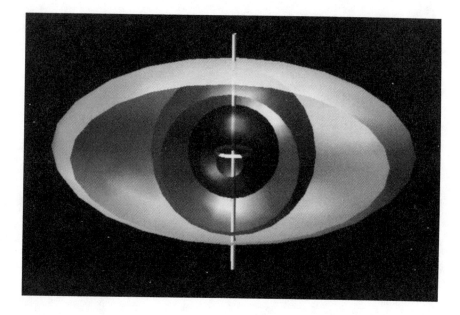

Three-dimensional representation of a rotating black hole. The outermost ellipsoid is the static limit. The two spheres correspond to the inner and outer event horizons. The ring at the center is the ring singularity. The vertical axis is the axis of rotation.

time appears to stop. The faster the hole spins, the larger the inner horizon grows until it merges with the outer horizon. If it were to spin any faster, both horizons would disappear, leaving a naked singularity."

Mr. Plex continues to cut. "Sir, I think I've finally hit the central singularity, but it feels like—like a ring." He cuts further. "*Mon Dieu!* It's a ring singularity, not a point."

"You're right. Rotating black holes have a doughnut-shaped singularity. Someday I'll tell you more about how this ring singularity might be used to travel to other universes. It may be possible to go through the ring without encountering infinitely curved space-time."

Mr. Plex opens his eyes wide. "Sir, can we get the computer to draw a cross-section of a rotating black hole?"

"Sure, go ahead."

Mr. Plex withdraws a battered notebook computer from under your bed pillow and types. In a few minutes, he hands you a print of a rotating black hole cross-section. "Sir, I've programmed the appropriate ellipses and circles."

"Very nice. The static limit is an ellipsoid because of the spinning of space, and the outer event horizon touches the static limit at the poles." You pause when you hear a crunching sound. "Mr. Plex! What on Earth are you doing?"

Mr. Plex is chewing on the discarded rind—the event horizon—from the static black hole. "Sir, I couldn't help myself. This is all very exciting, and I was quite hungry."

"Mr. Plex, when will you learn to behave like a respectful student?" Mr. Plex does not answer but simply continues to chew with his diamond teeth. A piece of the tattered event horizon drops to the floor.

The dust speck representing the central singularity floats away from the rind in Mr. Plex's mouth. The speck sparkles in the incandescent light of your cabin as it is carried upon the wisps and eddies of wind. *If only this singularity model were real,* you think. To gaze at the singularity would be a special treat indeed, because the only way to see one is to get so close that it would be impossible to counteract its massive, crushing gravitational pull.[5] Will humans ever gaze upon a singularity, a point where God crushes space and time into a cosmic molasses?

The speck drifts haphazardly around the room and finally comes to rest in your right ear, and for the briefest of moments you think you hear the chanting of monks.

THE SCIENCE BEHIND THE SCIENCE FICTION

Black-Hole Terminology

The lessons learned in your black-hole anatomy class are all valid and conform to what we presently know about the various classes of black holes. Astronomers have traditionally used different names to denote the different possible black-hole anatomies, and I list them here so you will be able to more easily follow the astrophysical literature:

1. *Schwarzschild black hole*—a static (nonrotating), noncharged black hole
2. *Reissner-Nordström black hole*—a nonrotating, charged black hole
3. *Kerr black hole*—a rotating, noncharged black hole
4. *Kerr-Newman black hole*—a rotating and charged black hole

"A Black Hole Has No Hair"

The weird expression "A black hole has no hair" is used by astronomers to say that all black holes have a curious kind of uniformity and that all the properties of a black hole can be summarized using just three quantities: mass, charge, and angular momentum (spin). All other information about a black hole prior to its formation is lost when matter is crushed into the singularity. For example, if you had a sufficient number of television sets crammed into the appropriate volume of space, the black hole formed would give you no clue that the original matter came from television sets. No amount of detective work would force a black hole to reveal the secret of its background. If probed, black holes can tell you only their mass, charge, and angular momentum. They could have been formed from bones, brains, or batons, but this private past remains shrouded in mystery.[6]

Fact File

Here are some interesting facts regarding spherical black holes.

- The interior of a black hole is empty, with a singularity at the center.
- A black hole's mass is crammed within the singularity, which essentially has a mathematical volume of zero.
- It is impossible to remain immobile in the interior of a black hole.
- If you are in a black hole, the only allowed trajectories are focused toward the central singularity.

- Space becomes timelike in the sense that it is constrained in one direction. The space coordinate always diminishes, just like outside the black hole, where all events move toward increasing times.

CHAPTER 8

And yet what can there be beyond the quantum mechanical wavefunction that may someday be written down to describe a multiverse in which the electron takes every possible path? Newton's laboratory table, perhaps, on which our multiverse sits enclosed in a crystalline sphere, dreaming that it is everything.

— George Zebrowski, *OMNI*, 1994

Embedding Diagrams for Warped Space-Time

"Sir, could you get that thing off of me?"

"It's just a bird, Mr. Plex. Nothing to get upset about."

The scolex suspiciously eyes the strange, hawklike bird that has alighted on his forearm. "What's it called?" The colorful bird flies off to the tallest tree in the aviary, where it begins to settle into its deep nest.

You turn to Mr. Plex. "It's called a hawking. It's attracted to bright lights. Maybe it was attracted to the glitter of your diamond body."

Mr. Plex gazes into the tree branches. "Its nest reminds me of a bowl shape, a paraboloid."

"Very good, Mr. Plex."

"Good?"

"That's the subject we will review today."

The scolex looks deeply into your eyes as his forelimbs quiver. "What do bowls have to do with black holes?"

"Come, Mr. Plex, sit down with me." You motion to a bench beside some ferns. "Wherever there's a mass in space, it warps space. Think of a bowling ball sinking into a rubber sheet. It's a convenient way to visualize what stars do to the fabric of the universe. If you were to place a marble into the depression formed by the stretched rubber sheet and give the marble a sideways push, it would orbit the bowling ball for a while, like a planet orbiting the sun." The scolex nods. "If I rolled a marble on the sheet far from the bowling ball—where there is no funnel in the rubber sheet—the marble wouldn't be affected. If I were too close, it would get sucked into the dent in space. The bowling ball warps the rubber sheet just like a star warps space. Far from the gravitating body, the curvature of the space 'funnel' is less pronounced."

The scolex smiles, revealing row upon row of diamond teeth. "Matter makes space bend. Space tells matter how to move."

You back away slightly. "You are quite astute today, Mr. Plex. Even light rays get bent by the curvature of space. Because of this, stars are sometimes not where they appear to be when you look at them in the sky."

"Can we study this space curvature on our computer?"

You nod and pull out a notebook computer from a small receptacle beneath the fronds of a palm tree. "Here are the equations describing how a mass, such as a star, curves space. The 'lift' of the funnel shape off the plane $z = 0$ is defined by two equations.[1] Inside the body of the star, we have

$$z(r) = \sqrt{\frac{R^3}{2M}} \left[\sqrt{1 - \left(1 - \frac{2Mr^2}{R^3}\right)} \right] \quad \text{for } r \leq R.$$

Outside the body of star we have

$$z(r) = \sqrt{\frac{R^3}{2M}} \left[\sqrt{1 - \left(1 - \frac{2M}{R}\right)} \right] + \sqrt{8M(r - 2M)} - \sqrt{8M(R - 2M)} \quad \text{for } r \geq R.$$

R is the radius of the star, M is the mass, and r is the radial vector used to draw the three-dimensional surface $z(r)$."

Mr. Plex scratches his head. "Sir, those equations look formidable." Faint vibrato plays in his voice.

"Not really. Outside the star, space is curved like a three-dimensional paraboloid of revolution. The interior looks like a sphere, and the interior and exterior geometries join together smoothly."

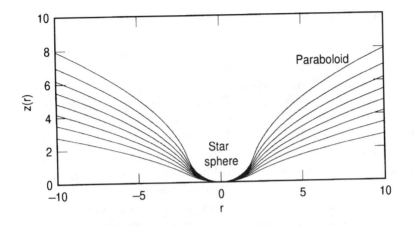

Two-dimensional representation of a gravitational curvature of space. The eight lines correspond to eight different star masses. The curve at the top corresponds to the heaviest mass, which causes the deepest depression. If you were to rotate the curves around $r = 0$ they would form paraboloids with spherical caps.

Mr. Scolex writes a computer program to draw the surface. "What do we pick for M and R?"

"Technically they're related by the density of the star, but we only want to get a qualitative feel for the shapes. Just be careful not to pick $R < 2M$ or we'll get a negative value under the square roots."

Mr. Plex presses a few buttons and gazes at the screen. "It looks like the nest of that hawking bird."

You nod and take a closer look at the computer screen. "Looks like the mouth of a trumpet to me." You pause. "These formulas assume that the star has uniform density. However, even in more realistic models, the surface looks qualitatively the same."

"Sir, what does this diagram mean for a point not on the trumpet shape?"

"Your funnel picture is only a convenient way to visualize the geometry of space around a star. It's called an embedding diagram. But only points lying on the 'rubber sheet' have physical significance. The three-dimensional regions outside the surface are physically meaningless."

A piece of wet bird guano falls from high up in the aviary and onto Mr. Plex's head as he says, "I think I can see where this is all leading. If we make the bowling ball more and more massive it will eventually distort the rubber sheet, representing space, so much that the mass will be cut off from the universe. It's as though the ball ripped a hole in the rubber sheet."

A bird cries high overhead, and you smell a strange odor. Perhaps it is time to bring the work crew in and clean out the aviary floor. "Yes, Mr. Plex—a black hole. If the mass is concentrated into a small enough space, it will stretch space so much that it's as though the mass pinches space off from the rest of the universe. Nothing can escape from the mass's tremendous gravitational pull."

"Sir, can we draw the embedding diagram for a black hole?"

"Sure. The curvature of space around a nonspinning black hole looks like the shape you'd get by twirling a parabola around its axis:

$$z(r) = \sqrt{8M(r - 2M)}$$

where M is the mass of the black hole. Can you write a computer program to draw it?"

"You bet," Mr. Plex says with a vague metallic twang. In another few minutes, the embedding diagram for a black hole appears on the screen of the notebook computer.

Mr. Plex points to the diagram. "Looks like the same funnel we saw for the star, except this one doesn't have a smooth spherical bottom."

"Far from the hole the surface is almost flat, which means the gravitational force is weak. The more the surface is curved, the stronger the gravitational force. When the wall is too steep, it will be impossible for anything thrown into the funnel to escape. This point of no return corresponds to the Schwarzs-

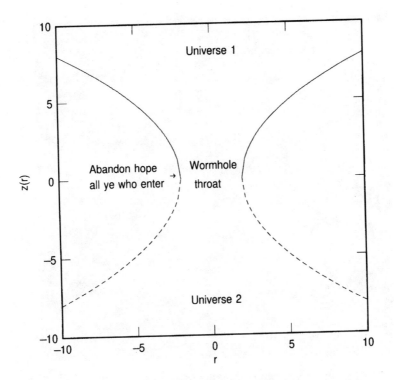

Cross-section of an embedding diagram for a static black hole. Although we often think of a black hole as creating a funnel shape, the equations give rise to the reflected curves (represented by dashed lines). A black hole is really a wormhole, or throat, connecting two universes.

child radius, also known as the gravitational radius or event horizon."

You reach over to the notebook computer and press a few buttons, and a mirror image of the funnel appears below the original one thereby creating a tube between two planes. "Mr. Plex, the Schwarzschild formulas have an extra version of the funnel built into the equations."

"Sir, I understand what the mouth of the top funnel means. But what does the bottom funnel imply?"

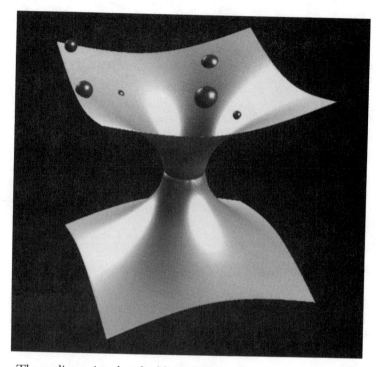

Three-dimensional embedding diagram for a static black hole. Objects on the surface of the embedding diagram represent bodies that are being "sucked into" the hole's hungry maw at the center.

"A black hole is really a wormhole connecting two universes." You motion toward the computer display. "This is called a Schwarzschild wormhole or Einstein-Rosen bridge. It connects two different regions of a single universe or perhaps one universe with another."

Mr. Plex's diamond eyes sparkle in the fading light of the aviary. The birds have grown quiet. "Sir, could one use a black hole to travel from one universe to another?"

"There are some practical problems with using nonspinning black holes for travel between universes, but there are other types of wormholes that might be useful for what you propose."

Mr. Plex's abdomen begins to rhythmically pulse. "Tell me!"

"Get a hold of yourself, Mr. Plex. Traversable wormholes are the subject for another day." You pause. There is a faraway look in your eyes. "In fact, I hope to travel though such a wormhole in a few days."

THE SCIENCE BEHIND THE SCIENCE FICTION

The embedding diagrams discussed by you and Mr. Plex give a useful intuitive feel for the gravitational effects of a mass on space. Far from the mass, where gravity is weak, there is little curvature. However, though these diagrams reveal the effects of gravity on the space portion of space-time, they don't say much about the time portion. As discussed in other chapters, general relativity predicts that gravity causes time to slow down. Watches tick more slowly in deep regions of space curvature than in flat regions.

C H A P T E R 9

If we wish to understand the nature of the Universe we have an inner hidden advantage: we are ourselves little portions of the universe and so carry the answer within us.

— Jacques Boivin, *Single Heart Field Theory*

When the universe has crushed him, man will still be nobler than that which kills him, because he knows that he is dying, and of its victory the universe knows nothing.

— B. Pascal, *Pensees, No. 347*

Gravitational Wave Recoil

"Sir, have you ever wanted to get married?"

You throw a stone into a pond and watch the concentric circular ripples begin to spread. If it weren't for the 30-foot-high ceiling, you'd think you were in a park rather than the ship's woodland sanctuary. You throw another stone. "Mr. Plex, I've considered marriage, but my passion for studying black holes has always gotten in the way."

The scolex considers this for a few seconds as he traces random patterns in the muddy shore. "Certainly a wife could have shared your interests. Take Mrs. Plex, for example. She's also a scientist of sorts."

You give him a noncommittal nod.

Mr. Plex reaches for a stone and throws it into the pond. Almost immediately, a small deep whirlpool forms in the center of the ripples. Mr. Plex wades into the water to get a closer look. "That's strange—"

"Back away, Mr. Plex. Perhaps a mini-wormhole has formed in the center of the ripples."

Mr. Plex suddenly turns and faces you, almost slipping in the pond. "What?"

You smile. "Just kidding, Mr. Plex. It was probably a thorne fish thinking your pebble was food and pulling it down. You better get out of there. Their spines can be deadly."

"Sir, I don't think they could do much to my diamond body."

The intricate play of light on the water is hypnotic. For a second you wish you were back on Earth gazing into a real pond beneath a blue sky. You shake your head to clear your thoughts. "The pond ripples remind me of gravitational waves from a black hole binary."

The scolex slowly emerges from the pond. "Binary?" His voice is eager.

"If pairs of black holes form close to one another they sometimes begin to orbit around one another. Each hole carves its own deep gravity well in the space-time terrain. These wells produce ripples of curvature that spread out with the speed of light."

Mr. Plex opens his eyes wide. "The holes spiral around one another?"

"Yes, the orbits are like tracks carved into the topology of space itself." You pause. "The gravitational waves that ripple outward push back on the holes just like a gun that recoils after shooting a bullet."

"Sir, the gravitational waves must force the holes to collide at some point."

You nod. "Quite right, Mr. Plex. The waves force the spiraling holes to get closer and closer. Finally their event horizons merge, and the two black holes coalesce to form one big spinning hole."

Mr. Plex grins. "Must be a sight to see."

"You wouldn't want to be too close. A lot of energy is released when they coalesce." You pause. "Can you guess how quickly two black holes must orbit one another?"

The scolex reaches into his abdomen for his notebook computer, and then he frowns. "I left my computer back at my room with Mrs. Plex. The battery needed charging."

"No problem."

You pull out a notebook computer from a receptacle hidden beneath a nearby tree stump and hand the computer to Mr. Plex. "The orbital period of two black holes traveling around one another is

$$2\pi\sqrt{D^3/(2GM_b)}$$

where D is their separation; G is Newton's gravitational constant, 1.327×10^{11} kilometers^3 per second^2 per solar mass; and M_b is the mass of each of the black holes."

Mr. Plex bangs furiously on the computer keys, his hairy, diamond forelimbs wreaking havoc with the plastic keys, and soon hands you a printout showing the orbital period as a function of distances between the centers of the holes:

Distance	Time
5000 kilometers	.9642038 seconds
15000 kilometers	5.010152 seconds
25000 kilometers	10.78012 seconds
35000 kilometers	17.8573 seconds
45000 kilometers	26.03351 seconds
55000 kilometers	35.17693 seconds
65000 kilometers	45.19434 seconds
75000 kilometers	56.0152 seconds
85000 kilometers	67.58376 seconds
95000 kilometers	79.85451 seconds

"Sir, I've assumed a mass of 20 solar masses for each hole. This means they are both 20 times as massive as your sun. The first result indicates that if the two holes are 5,000 kilometers apart, about the air distance from New York City to Los Angeles, they will orbit in a period of 0.9 seconds. That's fast!"

"Faster and faster, as they get closer."

"Sir, there's something I don't understand. How much time does it take for the holes to merge?"

"The gravitational-wave recoil squeezes them together, forcing them to spiral inward and coalesce:

$$t = \frac{5}{512} \times \frac{c^5}{G^3} \times \frac{D^4}{M_b^3}$$

where t is the time they will take to coalesce and c is the speed of light."

Mr. Plex types in the formulas and makes a table showing how much time is needed for the binary holes to coalesce for several different separations between the holes:

Distance	Time	
5000 kilometers	.009	days
15000 kilometers	.7413213	days
25000 kilometers	5.720071	days
35000 kilometers	21.97421	days
45000 kilometers	60.047	days
55000 kilometers	133.9961	days
65000 kilometers	261.3933	days
75000 kilometers	463.3254	days
85000 kilometers	764.3933	days
95000 kilometers	1192.712	days

"Sir, I'd bet that the embedding diagram would look beautiful for a binary system."

You smile. "Gorgeous."

Mr. Plex's claws open and close spasmodically. "Is it very difficult to compute and draw?"

"Not at all, if we just want a qualitative feeling for it. The intertwined double spiral of gravitational waves can be simulated with:

$$z = \cos\left(a\pi \times \sqrt{x^2 + y^2} + b \times \text{atan2}\,(y, x)\right).$$

You can fiddle with the parameters a and b to change various characteristics of the spirals. To make a realistic diagram, try $a = 10$ and $b = 2$. The atan2 is the arctangent function available in computer languages, and $(-1 \le x, y \le 1)$. To complete the representation, place two deep pits near the beginning of each spiral. These correspond to the two black holes." You pause. "For an even more realistic depiction, you can use the fact that the orbital radius depends on time as

$$r \propto (1 - t/\tau_0)^{1/4}$$

where τ_0 is the time it takes for the two holes to spiral together. Let's not worry about this for the time being, because we get an excellent qualitative diagram using the atan formulas."

Mr. Plex types some more on the keyboard and an embedding diagram with dual spirals and pits soon emerges. Mr. Plex holds up the image. "Beautiful!" He presses the Print button, and out comes a four-color, glossy print of the embedding diagram. "I'll have to bring this one back to my room to show Mrs. Plex."

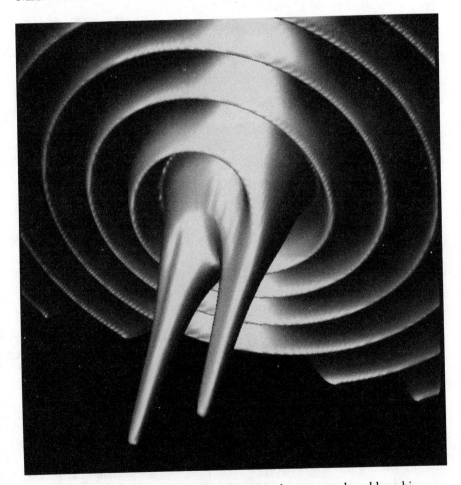

Embedding diagram showing gravitational waves produced by a binary system of two black holes. The two pits near the center are deep funnels in the fabric of space produced by the black holes.

You sigh. There's no one with whom you can share your hopes, visions, and fears. But perhaps that's for the best. Soon you will be attempting a very dangerous experiment involving the fabric of space-time. The single life is best for wormhole explorers.

Mr. Plex's diamond feet make squishy sounds in the mud. "Sir, did I say something wrong?"

"Everything's fine."

"You miss her, don't you?"

Your heart does a flip-flop. "Theano?"

"Yes. She—"

"Nothing lasts forever, Mr. Plex." Several thorne fish stick their bristly faces out of the water and seem to be watching you.

Could you possibly be jealous of the fact that Mr. Plex has a wife? Impossible. You imagine Mrs. Plex's mineral body, something you'd have no desire to hold in your arms. Could you share your intimate feelings with a scolex? You think about Mrs. Plex's bright diamond teeth, her twangy metallic voice, her multiple diamond forelimbs.

It's a good thing scolexes aren't made of flesh.

THE SCIENCE BEHIND THE SCIENCE FICTION

In a perfect universe, infinities turn back on themselves.
—George Zebrowski

Surely it heard me cry out—for at that moment, like two exploding white stars, the hands flashed open and the figure dropped back into the earth, back to the kingdom, older than ours, that calls the dark its home.
—T.E.D. Kline, *Children of the Kingdom*

The gravitational waves you observed with Mr. Plex are real.[1] All moving matter has a gravitational field that varies with time. The curvature of space fluctuates as a result of the changes in the spatial distribution of matter. These readjustments propagate, with the velocity of light, in the form of creases in space curvature as they travel across whatever background geometry exists in space. Collapsing stars with the least spherical shapes emit the most gravitational waves.

Kip Thorne has pointed out that the space-time curvature produced by the moon raises tides in the Earth's oceans, and the ripples of space-time curvature in gravitational waves from a black hole binary should similarly raise ocean tides. Unfortunately for physicists (but perhaps fortunately for life on Earth), the gravitational waves from coalescing black holes should produce tides in the Earth's oceans no larger than about 10^{-14} meters ($\frac{1}{10,000}$ the size of a single atom). Obviously, we can't measure such a tiny effect on our ocean's tides.

CHAPTER 10

He stopped, for saying the truth aloud was unendurable. He knew now why this tranquil life seemed like an after-life or a dream, unreal. It was because he knew in his heart that reality was empty: without life or warmth or color sound: without meaning. There were no heights or depths. All this lovely play of form and light and color on the sea and in the eyes of men, was no more than that: a playing of illusions on the shallow void.

— Ursula K. Le Guin, *The Farthest Shore*

Optical Appearance of a Collapsing Star

You pound on the door to Mr. Plex's room and begin to shout. "How would you like to see a collapsing star, Mr. Plex?"

In a few seconds, Mr. Plex appears at the door dressed in pajamas. You hadn't realized that scolexes wore pajamas.

The scolex steps out into the ship's corridor and closes the door behind him. "Sir, that would be a sight to behold," the scolex whispers. Perhaps he does not want to wake Mrs. Plex.

You smile. "I've got one on the viewscreen now."

The scolex closes the door. "Let's go."

As you walk through the corridors of the intergalactic zoo, you occasionally hear weird animal sounds coming through the doors to various habitats. You have never quite gotten used to the olfactory and auditory cacophony of the intergalactic zoo. Once on the ship's bridge, you motion to a dim light on the viewscreen.

"Mr. Plex, a little while ago I watched as this star imploded with ever-increasing speed, but now it has reached its relativistic stage."

"Explain."

"This star is in the last stages of collapse. It's almost shrunk to its gravitational radius. Notice that it almost seems frozen just outside its horizon."

Mr. Plex's eyes scintillate in the artificial light of the bridge. "Seems to be dimming."

"Yes, its total luminosity is decaying exponentially with time:

The cosmic dimmer-switch equation

$$L \propto \exp\left(\frac{-t}{3\sqrt{3}\,M}\right).$$

This equation gives the intensity of radiation received by distant observers. It's a 'continuum approximation' where one ignores the discreteness of photons." You pause. "The collapsing star appears to slow down as it gets closer to the gravitational radius. Remember, light from the star becomes redshifted, and clocks on the star's edge would appear to run more slowly."

You whip out a notebook computer from a receptacle in the wall and toss it to Mr. Plex. He catches it and begins to write a short program to produce exponential decay curves. "Sir, I'll make a plot of the light decay for several different star masses."

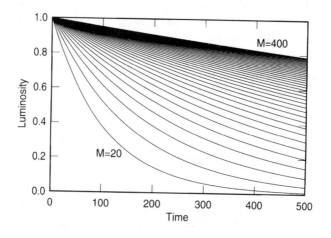

Various exponential decay curves for different star masses. Time units are arbitrary.

You nod. "Mr. Plex, the number of photons emitted before the star crosses its horizon is finite, so the exponential decay doesn't continue forever. As I said, the equation is an approximation that ignores the discrete photons emerging from the star."

"Okay, then do we know when the last photon escapes before the star crosses its gravitational radius?"

"Yes, the last photon emerges in

$$10^{-3}\left(\frac{M_{star}}{M_{sun}}\right)$$

seconds after the star starts dimming. Then the star is completely black."

The lights in the room begin to exponentially dim. The scolex looks up in confusion.

You begin to yawn. "Mr. Plex, it must be time to get back to sleep. Tomorrow is a big day for us."

THE SCIENCE BEHIND THE SCIENCE FICTION

> We wander as children through a cave; yet though the way be lost, we journey from the darkness to the light.
> —The Gospel According to Thomas (XV:1), quoted by T.E.D. Kline in Children of the Kingdom

As discussed in the chapter on gravitational time dilation, the dimming effect studied by you and Mr. Plex is real. The apparent luminosity of the star will decrease rapidly and the light will also be greatly redshifted, so rather than see the star slowly approach the Schwarzschild radius, very soon you would see no light at all coming from the body.

CHAPTER 11

He showed me a little thing, the quantity of a hazelnut, in the palm of my hand, and it was round as a ball. I looked thereupon with the eye of my understanding and thought: What may this be? And it was answered generally thus: It is all that is made.

— Julian of Norwich, 14th century

Gravitational Distension Near a Black Hole's Heart

"Sir, Mrs. Plex would like to talk to you."

Your heartbeat quickens. "What's this all about?"

The scolex looks down at his hindlimbs as they shuffle aimlessly in the dirt. "She heard that you were thinking of sending me into a black hole."

Before you have a chance to say anything, you hear footsteps approaching through the wildlife sanctuary. You brace yourself in anticipation of the meeting. A hawking glides high overhead, wings outstretched and motionless.

The leaves of some massive ferns part, and out steps Mrs. Plex. She looks directly into your eyes. "So you're the one who wants to send my husband into a black hole."

Your mouth hangs open. Mrs. Plex is not a scolex. Her body is not made of some hard mineral. Her voice has no metallic twang. Mrs. Plex is tall, blonde, elegantly attired in a fashionable yet no-nonsense business suit.

Mr. Plex seems to sense your hesitation. "Sir, I assume you have nothing against interspecies marriage?"

You turn to Mr. Plex. "She's human—"

"Sir, there's no need to talk down to her. She has two doctoral degrees."

"Mr. Plex, I never talk down to anyone." You turn and smile at Mrs. Plex.

Mrs. Plex comes closer. Her eyes narrow. "You seem to be wondering why I married Mr. Plex."

"Uh, well—"

"Haven't you heard that diamonds are a girl's best friend?"

You are speechless.

"Never mind. Just kidding." Her voice grows serious, almost confrontational. "So what's this I hear about you sending my husband into a black hole?"

"Mrs. Plex, you've got it all wrong. I just want him to get a little *closer* to the hole."

She puts her hands on her hips. "If he were to get sucked into a black hole by crossing its event horizon, is it true he would be pulled into the singularity at its center, no matter how hard he blasted his arterial rockets?"

"That's true. He would reach the singularity and be killed in

$$1.54 \times 10^{-5} \frac{M_{hole}}{M_{sun}}$$

seconds."

You reach under a tree stump, pick up a notebook computer, and toss it to Mr. Plex. "Mr. Plex, how quickly would you die?"

Mr. Plex writes a program that displays the time it would take him to die for various masses of black holes. He turns and hands you a computer printout.

Hole Mass (solar masses)	Time Remaining before Death (seconds)
20	3.080000E-04
70	1.078000E-03
120	1.848000E-03
170	2.618000E-03
220	3.388000E-03
270	4.157998E-03
320	4.928000E-03
370	5.697999E-03
420	6.467998E-03
470	7.237997E-03

"Sir, for a black hole of 300 solar masses, I would encounter the singularity in about 0.005 seconds."

"Nice work, Mr. Plex." You turn to Mrs. Plex. "There is one good thing."

Mrs. Plex is not smiling. "Yes?"

"Even though he can't escape or send messages to us once he has pierced the event horizon, he can still *receive* messages from us. We could even send food. The laws of physics only prevent things from coming *out* of the event horizon."

Mrs. Plex appears to be clenching her fists. "What good is that if he'll die in 0.005 seconds?"

"Ah, for maximum longevity, we may want to use a hole as large as those that inhabit the bright quasars: 10 billion solar masses. Mr. Plex could survive for hours before slamming into the singularity!"

Mr. Plex interrupts. "Sir, what happens if I ride on the surface of a collapsing star? My body can withstand the heat."

"Mr. Plex, your body would start to get stretched, as we discussed before. There'd be more force on your feet than on your head."

"Gravitational tidal forces?"

"Right. In fact we can compute the length of your distended body as the star collapses:

$$l_{body} = \sqrt{2M/R}\,(t_h - t_f) \propto R^{-1/2} \propto (\tau_{collapse} - \tau)^{-1/3}$$

where t_h and t_f are time coordinates along which your head and feet travel ($t_h > t_f$). R is the radius of the collapsing star, and τ is the time you would measure (if you were still alive) by a clock attached to your belt—which would be damn close to the time measured at your feet or head until your body obscenely elongates and the distance between your head and feet becomes comparable to the distance between you and the singularity. $\tau_{collapse}$ is the time at which you'd hit the $R = 0$ singularity." You sigh. "Mr. Plex, it would be wondrous to become one with a singularity."

Mrs. Plex appears nervous.

"Incidentally, Mr. Plex, the volume of your body actually decreases according to $V \propto R^{3/2} \propto (\tau_{collapse} - \tau)^{1/3}$.

"Sir, what does it all mean?"

"It means that your head-to-foot stretching increases to infinity, while your volume is crushed to zero."

"Sir, the equations seem to indicate that my length will increase to infinity, yet my head won't stick out of the event horizon as my feet rest on the singularity."

"Quite right, Mr. Plex. The extreme curvature of space near the singularity permits you to become infinitely long without shoving your head outside the horizon. Your head will never pierce the horizon again." You pause. "Finally your squashed corpse would merge with the quantum foam within the singularity."[1]

Mr. Plex's eyebrows raise. "Foam?"

"That's a topic we can discuss later." You pause. "Sounds like something you'd want to try? Would you like to pierce the horizon and experience the foam at its center?"

Mrs. Plex picks up the notebook computer and tosses it at your head.

THE SCIENCE BEHIND THE SCIENCE FICTION

As you read in previous chapters, if you were approaching a 10 solar masses black hole with a radius of 30 kilometers, you would be killed long before you reached the horizon at an altitude of 400 kilometers. J.-P. Luminet suggests that at the horizon the stretching effect would be the same as if you were hanging from a girder of the Eiffel tower with the entire population of Paris suspended from your knees.

Interestingly, black-hole appetites are not limited to interstellar dust and debris, or even planets. These cannibalistic creatures sometimes even feed on their own kind. Large black holes may devour nearby smaller ones. Two black holes can simply merge to form a larger black hole. Dozens of black holes can unite, forming larger and larger holes without limit. Astrophysicist Stephen Hawking has proved that if two black holes unite, the surface area of the final black hole must exceed the sum of the surface areas of the initial black holes. For these reasons the total black-hole portion of the universe is ever increasing.

C H A P T E R 12

Consider the true picture. Think of myriads of tiny bubbles, very sparsely scattered, rising through a vast black sea. We rule some of the bubbles. Of the waters we know nothing . . .

— Niven and Pournelle, *The Mote in God's Eye*

Quantum Foam

Mrs. Plex's eyes dilate with fear, or perhaps loathing. "Get off of me, you oaf."

You look closer to determine the cause of her outburst. There is a huge Ganymedean fractal spider climbing on her shoulder. From your zoology classes you know it has legs upon legs upon legs, each set of legs becoming smaller and smaller, until the smallest legs are only 10 atoms in length. The spider is said to be poisonous.

Mr. Plex, although nearby, seems paralyzed with fear. His right forelimb quivers periodically.

It's at times like this you're happy you have a black belt in karate. Without hesitation you leap through the air and launch your foot at the creature, instantly destroying it.

Unfortunately, Mrs. Plex is propelled backward and onto the ground. The ground is wet, and she lands in the mud with a squishy sound. As though this is not enough, she is made additionally uncomfortable by sluglike creatures that emerge from the mud. The creatures are called navanax, and they emit an odor of creosol and lime. Their cloaks of colorful flesh make them look like undulating, psychedelic jelly rolls.

Mrs. Plex looks up at her husband. "Help me out of this gross mess."

Mr. Plex extends his right forelimb.

You push it out of the way. "I can reach her, Mr. Plex." You graciously offer your hand and pull her out of the mud.

Mrs. Plex begins to wipe the mess from her skirt. She turns to you. "Thanks, I guess."

"Any time, ma'am." You pause. "It's not easy getting used to the wildlife here."

Mr. Plex steps forward. "Sir, there's something I've always wondered. Why are you head of an intergalactic zoo if your passion is black holes?"

"Good question, Mr. Plex. The job pays well. Anyway, I have the opportunity to take the ship to interesting astronomical areas and can study black holes in my spare time."

There is a movement near your foot. You look down and see the remains of the Ganymedean fractal spider. They begin to twitch and then foam. Mrs. Plex wrinkles her nose and backs away. You look closer. "Quantum foam, Mr. Plex."

Mr. Plex looks more closely at the foam. "What did you say, sir?"

"It reminds me of the quantum foam at the heart of a black hole."

Mr. Plex's eyes begin to dilate in delight. "I sense another lecture coming."

Mrs. Plex sits down on a tree stump and suspiciously eyes you.

You lower your voice and assume a professorial tone. "If you recall, last time we met we discussed the hypothetical results of Mr. Plex plunging into a black hole and traveling close to the singularity at its center. At the singularity, all the black hole's mass is squeezed into a region of space billions of times smaller than an atomic nucleus."

Mrs. Plex stretches her legs. "I'm glad you used the word *hypothetical* this time."

You are about to tell her not to interrupt your lecture, but then think better of it. You take a deep breath as a shiver runs up your spine. "Picture this: Your body parts have been thoroughly distorted by the huge tidal gravity within the black hole, and you are about to contact the singularity. The tidal gravity is so large that it completely deforms all objects in about 10^{-43} seconds, and quantum gravity takes over."

Mrs. Plex responds in a husky voice. "10^{-43} seconds?"

You nod. "Yes. 10^{-43} is called the Planck-Wheeler time approximated by $\sqrt{Gh/c^5}$, where

G (6.67×10^{-8}dyne–centimeters^2/gram^2) is Newton's gravitational constant, b (1.055×10^{-27}erg–second) is Planck's constant, and c (2.998×10^{10}centimeter/second) is the speed of light."

Is that a trace of admiration you see in Mrs. Plex's eyes?

Mr. Plex comes closer and glances from his wife to you. "What happens next as I approach the singularity?"

"You become one with the foam, Mr. Plex. Quantum gravity rips space and time from one another, and then destroys time as a concept and destroys the definiteness of space."

"Sir, you're saying that time doesn't exist inside a singularity?"

"Correct, Mr. Plex. And space becomes a 'seething,' probabilistic froth—a cosmological milkshake of sorts."

Mrs. Plex runs her fingers lingeringly through her hair. "Now, this sounds quite interesting. What do we know about the quantum froth?"

Your heartbeat increases in frequency and amplitude. "In the froth, space doesn't have a definite structure. It has various probabilities for different shapes and curvatures. It might have a 50 percent chance of being in one shape, a 10 percent chance of being in another, and a 40 percent chance of being in a third form. Because any structure is possible inside the singularity, we say the singularity is constructed from probabilistic foam, or quantum foam. Quantum gravity governs the probabilities for the various foam structures." You pause. "We believe quantum gravity also determines the probabilities for the singularity giving birth to 'new universes'—new regions of space-time."

Mrs. Plex seems deep in concentration. "Just like the big bang singularity gave birth to our universe 15 billion years ago?"

You nod and smile. "Yes." You begin to pace. "Quantum foam is everywhere: in singularities, in space, even in your body. But you'd need a powerful microscope to examine it. We're talking size scales around 10^{-33} centimeters."

Mr. Plex's forearms begin to quiver as he looks at you. "Can we—?"

"Yes?"

"Can we simulate this on a computer?"

Before you have a chance to answer, Mrs. Plex stands up. "Yes, we can. Got a notebook computer around here?" She sticks her hand out to her husband.

Mr. Plex looks down at the mud. "Left mine in our room, dear."

A blush of pure pleasure rises to your cheeks as you pull a notebook computer from between the leaves of a fractal fern. You gently hand it to Mrs. Plex.

She begins to furiously type using her slender fingers. In a few minutes a probabilistic froth undulates on the screen.

Two-dimensional computer model of a quantum foam embedding diagram. The algorithm produces seething probabilistic froth and tunnels. Quantum foam is thought to exist within a black hole's singularity where the geometry and topology of space are probabilistic.

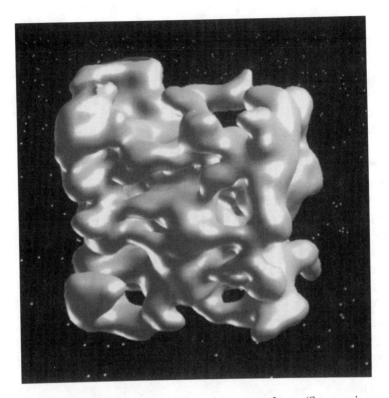

Three-dimensional representation of quantum foam. (See previous figure legend for details.)

Mrs. Plex looks deeply into your eyes. "This three-dimensional froth even produces tunnels and wormlike tubes commonly depicted in embedding diagrams for quantum froth. The connecting bridges in the foam correspond to wormholes between different universes or between different places in the same universe."

Your mouth hangs open. "How in hell did you make such beautiful shapes?"

"Secret algorithm from the twentieth century."

You pout. "Please, can you tell me?"

She smiles. "It's based on a cellular automaton consisting of a grid of cells that can exist in two states: occupied or unoccupied. The occupancy of one cell is determined from a simple mathematical analysis of the occupancy of neighbor cells." She pauses as she stretches her arms. "Though the rules are simple, the patterns are very complicated and sometimes seem almost random, like a turbulent fluid flow or the output of a cryptographic system."

Mrs. Plex presses a button on the notebook computer and hands you a listing of the computer program. "This describes the rules in two dimensions, and I've done something similar to construct the three-dimensional embedding diagrams, which of course are much more realistic and interesting to look at." She pauses. "This is merely a nice recipe for making structures qualitatively quite similar to embedding diagrams for quantum foam and mimicking its probabilistic, unpredictable nature. No one really knows too much about the structure of quantum foam."

You give a low whistle because you are quite impressed with the depth of Mrs. Plex's knowledge and her facile manipulation of the computer. "Mrs. Plex, may we have dinner sometime? I'd like to discuss black-hole simulations further with you."

"Dinner?" she says.

A hawking flies overhead and time seems to stop. No one moves. The bird seems suspended, motionless. Reflections from Mr. Plex's body make you feel that you are standing on the periphery of some gigantic crystal as your image is many times reflected.

You feel a chill, an ambiguity, a creeping despair. Mr. and Mrs. Plex are still. Their eyes are bright, their smiles relentless and practiced. For a moment, the navanax in the mud seem to glow. But when you shake your head, the glow is gone. Just your imagination. But the navanax remain. Cruel. Nightmarish. You feel like you're caught in a maze and all the air is suddenly evacuated.

You shake your head again, and the world is back to normal. Perhaps you are just nervous about your plan to create and enter a wormhole to another universe. Perhaps you feel bad for asking Mrs. Plex to dinner in front of her husband.

You reach into your pocket and hand Mrs. Plex a card. "Should you ever need to reach me, I can always be contacted at this phone number." In addition to your phone number, your business card has a symbol of a black hole, and below your name are the words "Black Hole Specialist."

Mrs. Plex smiles and places the card in her hip pocket.

THE SCIENCE BEHIND THE SCIENCE FICTION

Of What Singularities Are Made

How long would it take you or Mr. Plex to become one with the singularity's quantum foam after crossing the event horizon? Would you have any time to explore the interior of a black hole before the crushing experience? As alluded to in chapter 3, it turns out that the more massive the hole, the more time you have for exploration. For a 10 solar masses black hole, you only have a thousandth of a second, but for a behemoth black hole in a galactic core, you should be able to explore for about an hour.

A space-time singularity is always produced by gravitational collapse. In fact, between 1965 and 1970, Roger Penrose and Stephen Hawking proved that, according to general relativity, there must be a singularity of infinite density and space-time curvature within all black holes. Hawking notes that at the singularity many laws of science and our ability to predict the future break down. However, if you remain outside the black hole, you would not be affected by this failure of predictability, because neither light nor any other signal could reach you from the singularity. This amazing fact led Roger Penrose to propose a cosmic censorship hypothesis paraphrased as, "God abhors a naked singularity." Hawking notes that the singularity produced by gravitational collapse occurs only in special places, like black holes, where it is "decently hidden from outside view" by an event horizon.

Graphical Simulation

The quotation from Niven and Pournelle's futuristic science fiction novel at the beginning of this chapter describes both the vast mystery of our universe and the strangely shaped objects we might one day encounter in outer space. In the 1990s, however, we don't need a spaceship to explore strange new worlds consisting of bubble-like forms. All that is required is a small set of mathematical algorithms running on a good graphics computer.

Quantum-foam embedding diagrams decorating living rooms of the future will include computerized versions where the probabilistic froths are simply mathematical/computer graphical entities displayed on a computer screen. To create the undulating froth, you may first wish to construct a two-dimensional model where the forms move, coalesce, and break up in the infinitely thin space between two glass plates. The simulation involves the use of "cellular automata." Cellular automata (CA) are a class of simple mathematical systems that are becoming important as models for a variety of physical processes. CA are mathematical idealizations of physical systems in which space and time are discrete. Usually cellular automata consist of a grid of cells that can exist in two states: occupied or unoccupied. The occupancy of one cell is determined from a simple mathematical analysis of the occupancy of neighbor cells. One popular set of rules is set forth in what has become known as the game of "Life." Though the rules governing the creation of cellular automata are simple, the patterns they produce are very complicated and sometimes seem almost random, like a turbulent fluid flow or the output of a cryptographic system.

To create a CA, each cell of the array must be in one of the allowed states. The rules that determine how the states of its cells change with time are what determine the cellular automata's behavior. There are an infinite number of possible cellular automata, each like a checkerboard world. The shapes in this chapter were produced by initially filling the CA array with random 1s and 0s. The rules of growth that determine the states of the cells in subsequent generations are discussed in the next section.

The systems in this chapter evolve in discrete time according to a local law. As with most cellular automata, the value taken by a cell at time $t+1$ is determined by the values assumed at time t by the neighboring sites and by the considered site itself:

$$c_{i,j}^{t+1} = f(\, c_{(i,j)}^t,\ c_{(i+1,j+1)}^t,\ c_{(i-1,j-1)}^t,\ c_{(i-1,j+1)}^t,\ c_{(i+1,j-1)}^t,\ c_{(i+1,j)}^t,\ c_{(i-1,j)}^t,$$
$$c_{(i,j+1)}^t,\ c_{(i,j-1)}^t).$$

In this equation, $c_{i,j}^t$ denotes the state occupied at time t by the site (i, j). The nine-cell template used in this chapter is referred to as the Moore neighborhood (as opposed to the von Neumann neighborhood consisting only of orthogonally adjacent neighbors). One interesting simulation simply examines the neighbor sites to determine whether the majority of neighbors are in state 1. If so, then the center site also becomes 1. We can represent those cells in the on (1) state as black dots on a graphics screen. In other words, this rule is a voting rule that assigns 0 or 1 according to the "popularity of these states in the neighborhood," and interestingly it generates behavior found in real physical systems. This simple majority-rule automata produces hundreds of coalesced, convex-shaped black areas but does not lead to interesting graphical forms. A way to destabilize the interface between 1 and 0 areas is to modify the rules slightly so a cell is on if the sum of the 1-sites in the Moore neighborhood is either 4, 6, 7, 8, or 9; otherwise the site is turned off. This rule has been studied previously (in lower resolution), and, because it uses a Moore neighborhood, it was termed M46789 by Vichniac in 1986. Such simulations have relevance to percolation and surface-tension studies of liquids.

The quantum foam in the first figure Mrs. Plex produced shows an M46789 2-D form that has evolved after several hundred time steps from random initial conditions on a 2,000-by-2,000 square lattice. The tiny dust specks sparsely scattered between the coalescing blobs are not dirt left by the graphics printer, but rather they are stable structures such as

★★
★★

where each site has exactly 4 "on" members (out of 9) in its neighborhood, and thus stays on. The surrounding sites that are off have at most 1 or 2 neighbors that are on, and thus they stay off. There are probably quite a few other stable structures like this, though this rule does not seem to give rise to the zoo of stable objects allowed by, say, the game of Life.

The twisted-majority rule for 2-D Moore neighborhoods just described naturally leads to curiosity about how 3-D foam forms evolve when displayed using high-resolution graphics on graphics supercomputers (see the second figure Mrs. Plex produced). I have recently created 3-D foam simulations by extending the 2-D simulation to an $M(13, 15, 16, 17, 18, 19, 20, 21, 22, 23, 24, 25, 26, 27)$ simulation on 50-by-50-by-50 and 100-by-100-by-100 3-D lattices. For the remainder of this chapter I will call this rule $M14$ to indicate where the twist, or skip, in the majority rule takes place.

To create the second figure, $M14$ foam is allowed to evolve for about 50 generations starting from randomly filled arrays of 0s and 1s. To visually decrease grid artifacts, the final site values for the cellular automata are determined by replacing the center site value with the mean value of its 27-cell Moore neighborhood prior to display. This effectively turns the discrete on/off collection of sites into a slightly smoother continuum. The final graphic form is created by computing an equal-valued surface, or "isosurface," at a particular value in the 3-D data set. These equivalued surfaces, which represent the smooth boundary between the 1s and 0s, are the final form the viewer actually sees. The second figure was rendered using a program that computes and represents surfaces as a collection of small triangular facets. The triangles that make up the surface are smoothed, shaded, rotated, and lighted using a general-purpose display program.

As a submarine pilot might explore coral formations in the Sargasso Sea, computer graphics and powerful computers allow one to explore the strange and colorful $M14$ quantum froth tunnels and caverns using a mouse. As the simulation progresses, starting from randomly mixed 0s and 1s, those cells in the foam (that are in the 1 state) move around randomly until they meet and join by cohesion—forming visually interesting aggregates due to the $M14$ rule.

Note that none of the algorithms I have just described is based on quantum gravity. The computational recipes are simply meant to produce seething probabilistic froth and tunnels reminiscent of quantum-foam embedding diagrams.

C H A P T E R 13

I saw Eternity the other night,
Like a great Ring of pure and endless light,
All calm, as it was bright;
And round beneath it, Time in hours, days, years,
Driven by the spheres
Like a vast shadow moved; in which the world
And all her rain were hurled.

— Henry Vaughan (1622–1695), "Eternity"

Black-Hole Recreations

Mr. Plex watches a scolex child playing by some nearby rocks. He points her out to you. "Diamond girl."

You notice the light from her body sparkling.

"Sure does shine."

"That's my cousin, sir."

Mr. Plex turns to her and calls. "Come here, Natasha."

Natasha, a small, glimmering scolex, slowly undulates toward you with her multiple legs. You pat her gently on the head, and then she runs off and plays by the edge of a simulated sea. Within the ocean, Ganymedean jellyfish, ctenephores, and all manner of Silurian sea creatures frolic and prance, their diaphanous tentacles and evanescent egg sacs floating close behind in the cold liquid.

Mr. Plex yawns. "Had some trouble getting to sleep last night, sir." He pauses. "Sir, I'm worried about your proposed trip inside a black hole or wormhole."

"Mr. Plex, did you try counting sheep? It works for me when I have trouble getting to sleep."

"Counting sheep?"

You look at Mr. Plex. Do scolexes dream of crystal sheep?

"Sheep?" he repeats.

"Never mind, Mr. Plex. Anyway, we're here in the seashore habitat for a little rest and relaxation. I'm delaying my adventure for another few days." You pause. "Care to play a game?"

The scolex nods. "Certainly."

Mrs. Plex wanders over. You watch as she brushes back some hair that the faint wind has caressed out of place. "May I play?" she asks.

93

You nod. "It's called the Black Hole Game. You can play it with a pencil and paper on a graph paper that is used as a dot grid. First make an array of 26-by-26 dots." You pause. "The game is played with 'Ganymedean worms.' Each worm consists of five connected lines that stretch over six dots. One of the dots at the end of the worm's body is circled to indicate its head. The worms can contort their bodies at right angles to form different shapes, but they can't re-use a grid point through which another part of their body passes."

Mrs. Plex stares intently at a piece of graph paper you have in your hand. "Each worm body passes through six dots?"

"Correct. And the allowed movements are up, down, right, and left. The worm's body may not lie along a diagonal. Here are some examples." You sketch on the paper.

```
                     -
1. Hat        _| |_

                     --
2. Dipper     |_|

                    -
                   _|
3. Steps    _|

4. Line     -----

5. L-shape  __|
```

Mr. Plex traces some of the figures in the sand with his forelimb. "I wonder how many different worm shapes you could make from five connected line segments."

You furiously sketch on the graph paper. "There are 21 'Palmer' configurations:

Worm Contortion Patterns

Category 1: Twelve with a "|_" at the end (The Slaves)

```
                                        |
                _    |    _    |    _|   _
   _   _   _   _   |_   |_   |_   |_  |_  |_  |_  |_
 |_ |_| | |  | | |    |    |    |   |   |   |   |   |
 |
```

Category 2: Six with a "I_I" at the end (The Peasants)

```
 _    _      _        _
| |  | |    | |   _| |   _| |  |_| |
|_    |    _|     |
   |
```

Category 3: Four with "--" at the end (The Lords)

```
                    |
              |     |
       _   _  |_    |
      | |  |     | |
      | |  |     | |
```

"For example, the Lords cannot have a hook, such as _I, at their ends." You pause, noting the admiration in Mrs. Plex's eyes in response to your intellectual prowess. "In the game, each player, on his or her turn, has to position a worm on the grid."

Mr. Plex's abdomen pulsates. "What are the rules?"

"The worms of two players cannot overlap or share the same dot. In other words, every dot in the grid can only be occupied once. The worms cannot lie across one another but can be tightly folded and intertwined. The game is over when no one can add another worm to the grid. To determine who wins, count the number of Lords, Peasants, and Slaves. Each Lord is worth 3 points; each Peasant, 2 points; and each Slave, 1 point."

Mrs. Plex grins. "Let's try it."

You smile back. "To start a worm, select a dot and then move to an adjacent dot, either right, left, up, or down. Repeat this four more times." You pause and draw a worm. "I'll use an open circle on the end of my worm to denote its head. You use a closed circle." You pause. "We take turns filling the grid with worms."

Mrs. Plex runs her fingers through her blond hair. This motion is like an angel fish swimming through water. "How does this relate to black holes?"

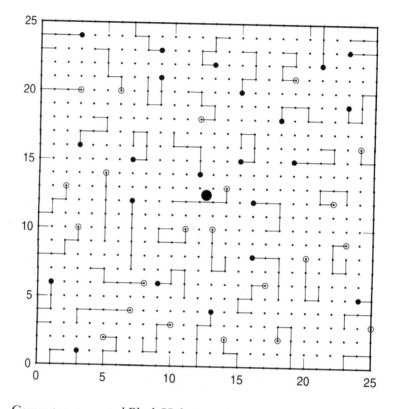

Computer-generated Black Hole game board.

"Ah, good question. I forgot two last constraints. There is a black hole at the center of the grid denoted by a big black dot. The worms, as they float through space, can sense gravitational fields and so naturally point their heads toward the black hole. Therefore, the head of each worm must be closer to the hole than the tail of each worm. Additionally, the worms hate one another, so their heads may not be on adjacent, orthogonal grid units."

Mrs. Plex reaches for your pen. "Okay, let's play."

You play for several minutes. Mrs. Plex seems to prefer playing near the corners. You prefer to position your worms close to the black hole.

You look into Mrs. Plex's eyes. "On Ganymede the great philosophers study this game for days, and tournaments last for weeks."

Mrs. Plex returns your stare. "Interesting."

After a few more moves, the board is so filled with worms that you can't add another worm without crossing over a worm on the board. Both you and Mrs. Plex add up the points for each worm.

You take a step back. "Mrs. Plex, you won! Are you sure you've never played this game?"

Mrs. Plex grins. "Quite certain."

Mr. Plex comes closer. "I wonder what the best strategy is. Does the first player have an advantage? I wonder how many people can play on the same board."

You shrug. "All unanswered questions, Mr. Plex. I once saw three people playing using different colored worms."

Mr. Plex's forearms begin to quiver. "Sir, I think I'll write a computer program to play itself."

"Good idea, Mr. Plex. That will give me some time to get to know your wife and explain to her all the experiments we've been doing."

Mr. Plex nods. Next he extends of one of his forelimbs into the sand, withdraws a notebook computer, and then begins to type. "I'll write a program so the computer randomly creates worms and then checks to see if they do not travel along a previously occupied grid point. I'll also make sure their heads point to the black hole at the center of the board."

"Excellent, Mr. Plex."

From the corner of your eye, you sense a movement. The diamond girl waves to you, and you wave back. Mrs. Plex is wading in the ocean.

Soon something resembling a green and red Möbius ribbon with blue eyes and a pompadour of greasy hair comes out of the depths and wriggles closer to Mr. Plex. Jellyfish creatures with purple bladders float to the surface of the amber sea. Seconds later a vast purply mass, perhaps their mother, joins the jellyfish. Its myriad fractal arms curl and twist like a nest of boa constrictors.

Life is good, you think to yourself. The sand. The surf. The sea. There's nothing like a day at the beach.

THE SCIENCE BEHIND THE SCIENCE FICTION

I invented the Black Hole Game described in this chapter and received a few interesting comments from players on the Internet. For example, Ed Pegg, Jr. (a cruciverbalist from Colorado and former computer programmer at NORAD, where he tested his programs by running nuclear attack scenarios), wrote to me:

> I've done some tinkering with the Black Hole Game, and think that one more rule needs to be added. The two types of worms (X and O) don't like one another's odor, and don't trust each other very much, thus the new rule: The head of a worm may never be adjacent to the body of an opposing worm. This adds a territorial flavor to the game, especially since all heads must be aligned toward the black hole. Without it, every move hurts you as much as it hurts your opponent. With the new rule, you can hurt your opponent without hurting yourself. For example, if the first player plays the worm a3, a4, a5, a6, a7, a8, the second player cannot play the worm b1, b2, b3, b4, b5, b6, since the head (b6) would "touch" the body of the opposing worm (at a6) (see following diagram). However, the move b1, b2, b3, b4, b5, c5 would be fine.

```
   1 2 3 4 5 6 7 8
a     X X X X X
b 0 0 0 0 0
c         0
d
```

The strategy now becomes one of territorial control. b2-b7 would be a good opening move, since a later move of a1-a6 is almost assured.

Unanswered Questions

I would be interested in hearing from readers who have experimented with the black-hole game, either using a pencil and paper or using a computer. Here are some interesting variations:

- *Lords only.* In this version, you only play with Lords. In other words, you can only use a shape that is one of the four Lords. How does this change the game strategy?
- *No edges.* In this version, you play on a game board where the top edge "connects" to the bottom edge, and the right edge connects to the left edge. (Think of wrapping the board onto a doughnut.) How does this change the game strategy?
- *Hexagonal board.* How would the game strategy change if five-segment worms were placed on a hexagonal board with a hexagonal packing of grid points, rather than on the standard square board?
- *Fibonacci Worm game.* This game involves the use of the Fibonacci sequence to determine the number of moves a player is permitted during his or her turn. (Fibonacci numbers are such that, after the first two, every number in the sequence equals the sum of the two previous numbers. For example, 1, 1, 2, 3, 5, 8, . . .) To play *Fibonacci Worm*, you make one move, then your opponent makes a move. You make two moves, then your opponent makes three moves. You make five moves, and so on. Strategy in this game would seem difficult because the next player gets so many more moves than you do.

C H A P T E R 14

The earth, that is sufficient,
I do not want the constellations any nearer,
I know they are very well where they are,
I know they suffice for those who belong to them.

    — Walt Whitman, *Leaves of Grass*, 1855

The center of the black hearth, of setting suns on the shore: ah!
wells of magic.

    — Arthur Rimbaud, *Illuminations*

Mathematical Black Holes

"Sir, why are we sitting in the dark?"

"To heighten the suspense, Mr. Plex." You are in Mr. Plex's cabin. Mrs. Plex is taking a nap in the adjacent room. You flick on a light switch, revealing a notebook computer with a set of equations glowing on its screen. "Mr. Plex, I think you'd find mathematical black holes fascinating."

"Mathematical?"

"Yes, mathematical formulas that behave like black holes. You position points in space, and the equations suck them into the center. Beautiful to watch on a computer." You pause. "For example, consider the equation that governs an electrical circuit with a resistor, capacitor, and inductor. The differential equation looks like:

$$L\ddot{x} + R\dot{x} + \left(\frac{1}{C}\right)x = 0.$$

Here x represents the charge on the capacitor, $\dot{x}$ represents the current in the loop, and $\ddot{x}$ represents the change in current. R is the resistance; L, the inductance; and C, the capacitance."

Mr. Plex's forelimbs quiver. "Sir, differential equations scare me."

"No matter. We can easily implement this on a computer if we express it as follows:

$$x_t = x_{t-1} + \lambda y_{t-1}$$

$$y_t = y_{t-1} + \lambda \left[\left(\frac{-1}{LC}\right)x_{t-1} - \left(\frac{R}{L}\right)y_{t-1} \right]$$

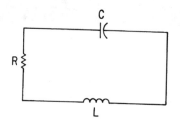

RLC circuit. A capacitor is a device for storing an electric charge on two conducting plates separated by an insulating material. The symbol for a capacitor resembles an "=" sign. A resistor, symbolized by a zigzag line, is an electronic component with a fixed amount of resistance to the flow of electrical current. An inductor is a coil of wire that can store energy in the form of a magnetic field. The symbol for an inductor is a little coil.

where $\lambda > 0$ is a constant known as the step size of the numerical solution. Let's keep λ small ($\lambda \sim 0.1$). Also, $-3 < x < 3$ and $-3 < y < 3$, and R, L, and C can be set to 1.0. This is a nonlinear rule that maps a point X_t to a new point X_{t+1} and can be thought of as a discrete-time dynamical system."

Mr. Plex shuffles over to the notebook computer and types a computer program. After a few seconds, a fantastic spiral image appears on the screen.

"Beautiful, Mr. Plex. Your graphics reveal that all initial points (x, y) have spiral trajectories that converge to the fixed point in the center of the figure. I like to think of the fixed point at $(0, 0)$ as a mathematical black hole. No particle can escape its deadly grip!"

Mr. Plex instructs the computer to continually add randomly positioned points and watches them spiral inward. "Sir, in a sense, all points are within the Schwarzschild radius of this mathematical black hole."

You nod.

Mr. Plex watches the shape dynamically unfold on the screen as it reveals the damped response of the RLC circuit. Occasionally the light from the computer screen is reflected from his body, producing an array of scintillating stars.

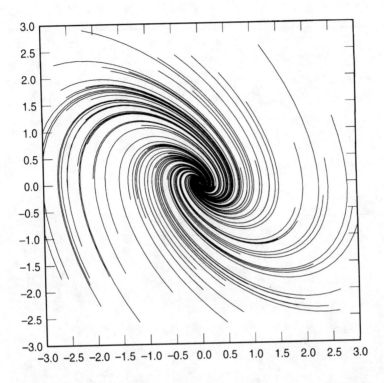

Phase portrait of an RLC circuit created by plotting trajectories through time. All initial points are inexorably pulled into the "mathematical black hole" at the center.

You cough to get Mr. Plex's attention. "The Mandelbrot set is also a mathematical black hole of sorts."

"Sir, you mean that famous bush-shaped fractal object?"

"Yes. You can create it by using mathematical feedback loops. Starting with $z = 0$, you continually iterate (repeat) $z = z^2 + c$ for complex numbers. Each point c that starts inside the Mandelbrot set can't escape. In fact, points in the main cardioid swirl like a whirlpool as they drain into a fixed point. Points outside the bushy shape fly off to infinity."

"Sir, the event horizon separating trapped and exploding points is quite intricately shaped."

Artistic rendering of the Mandelbrot set. All points outside the set
are free to travel to infinity as the equations are iterated in a mathematical feedback loop. The fates and trajectories of initial points inside the set, however, are forever bounded. It can be proved that all
starting points outside an "event horizon" defined by a circle of radius 2 centered at the origin fly off to infinity.

"Quite right. The event horizon is fractal." You pause. "I
also like the Julia sets, close cousins of the Mandelbrot set.
Julia-set fractals are also created with $z = z^2 + c$, except c is held
constant while initial values for z change." You press a few but-

tons on the computer, and beautiful colorful images appear. "A particle that starts its life in the black region is forever confined in a prison with intricate walls, trapped, never knowing the freedom of its infinitely near brethren—brethren born outside the crinkly event horizon exploring far and wide, eventually exploding outward to meet their God, infinity."

There is admiration in Mr. Plex's eyes. "Sir, you're a poet."

Suddenly an alarm rings on the ship's intercom. A voice says, "Warning, the elephants have escaped from the African habitat."

"Let's go, Mr. Plex. Duty calls. We'll continue our lessons later."

An elephant suddenly crashes into the room and sits on Mr. Plex.

"Do not worry," he gasps. "My diamond body can take it."

"I am not worrying, Mr. Plex. I was just thinking about how black holes can evaporate."

Mr. Plex breathes rapidly as the elephant covers his entire body. The scolex's voice is hardly audible from beneath the gray mountain of flesh. "Evaporate, sir?"

You call back to him as you run from the room and begin to search for other escapees. "That, Mr. Plex, is the subject for our next lesson."

THE SCIENCE BEHIND THE SCIENCE FICTION

Explore the Electronic Black Hole

The RLC circuit you discussed with Mr. Plex is often studied by beginning physics students. As background, many electrical circuits have loops of wire. When a change occurs in a magnetic field that passes through the area enclosed by a wire loop, a current is caused to flow momentarily in the loop.

The actual circuit you discussed with Mr. Plex consists of a resistor R, capacitor C, and inductor L, all in series (the first figure above). The circuit has no voltage source in it. As a typical example, suppose a charge Q is placed on the capacitor while a switch in the circuit is open so no current can flow. If you close the switch, the charge will begin to flow off the capacitor and a current will start to flow. The current rises and energy becomes stored in the inductor coil.

At some point the capacitor loses all its charge, and its original energy store is lost. At that instant an equivalent energy is stored in the inductor. Because the inductor will not let the current drop to zero instantaneously, current will continue to flow and will charge the capacitor opposite to its original charge. The entire back-and-forth charging process will repeat itself, and the circuit will oscillate indefinitely with clockwise and counterclockwise current flows.

If resistance is present in the circuit, energy will be lost through heating. Like a marathon runner who slowly loses his wind and can no longer even walk, the current in the inductor (and charge on the capacitor) must eventually decay to zero.

Depending on the relative values of R, L, and C, the circuit can be made to behave in various wild ways, and the resulting graphics are always fascinating to watch. In the case you presented to Mr. Plex, the gradual decay of the circuit's oscillating current corresponded to the whirlpool decay to zero (Figure 14.1). There are other kinds of possible behaviors for an RLC circuit. Here is a summary. If $R = 0$, then the circuit oscillates in a sinusoidal fashion and does not decay. If $0 < R/2L < < \sqrt{1/LC}$, the circuit is "underdamped" and exhibits gradually decaying oscillations. Finally, if $R/2L = \sqrt{1/LC}$, the circuit is "critically damped" and the circuit decays to zero with no oscillations.

Explore the Fractal Black Hole

These days, computer-generated fractal patterns are everywhere. From squiggly designs on computer art posters to illustrations in the most serious of physics journals, interest continues to grow among scientists and, rather surprisingly, artists and designers. The word *fractal* was coined in 1975 by mathematician Benoit Mandelbrot to describe an intricate-looking set of curves, many of which were never seen before the advent of computers with their ability to quickly per-

FIGURE 14.1. RLC whirlpool. No starting points can escape the deadly grip of the "mathematical black hole" at the center. To produce this form, use the RLC equation provided in this chapter and use time to displace points in the z direction. Points at later times correspond to points at the bottom of the whirlpool.

form massive calculations. Fractals often exhibit self-similarity, which means various copies of an object can be found in the original object at smaller-size scales. The detail continues for many magnifications, like an endless nesting of Russian dolls within dolls. Some of these shapes exist only in abstract geometric space, but others can be used as models for complex natural objects such as coastlines and blood-vessel branching. Interestingly, fractals provide a useful framework for understanding chaotic processes and for performing image compression. The dazzling computer-generated images can be intoxicating, motivating students' interest in math more than any other mathematical discovery in the past century.

Since its discovery around 1980 by Benoit Mandelbrot, the Mandelbrot set has emerged as one of the most scintillating stars in the universe of popular mathematics and computer art. The set also has an important connection with stability and chaos in dynamical systems. The Mandelbrot set for the iterative process $z \to z^2 + c$ (for complex numbers) has been widely investigated in recent years. Briefly stated, the Mandelbrot set is defined as the set of complex values of c for which successive iterates of 0 under f_c do not converge to infinity.[1] The computer software listing will help you implement this using a computer. All points outside the set are free to travel to infinity as the equations are iterated in a mathematical feedback loop. The fates and trajectories of initial points inside the set, however, are forever bounded. Points in the main cardioid are sucked down into a fixed point within the main body. It can be proved that all starting points outside an "event horizon" defined by a circle of radius 2 centered at the origin fly off to infinity.

Other Mathematical "Black Holes"

It's like asking why Beethoven's Ninth Symphony is beautiful. If you don't see why, someone can't tell you. I know numbers are beautiful. If they aren't beautiful, nothing is.
—Paul Erdös

In addition to the mathematical systems described in the preceding sections, black-hole-like behavior can be produced by other "dynamical systems" that are also deep reservoirs for striking images. Dynamical sytems are models containing the rules describing the way some quantity undergoes a change through time. For example, the motion of planets around the sun can be modeled as a dynamical system in which the planets move according to Newton's laws. You can create intricate designs by tracking the behavior of mathematical expressions called differential equations, such as the one in this chapter used to model the behavior of *RLC* circuits. Think of a differential equation as a machine that takes in values for all the variables and then generates the new values at some later time. Just as one can track the path of a jet by the smoke path it leaves behind, computer graphics provides a way to follow paths of particles whose motion is determined by simple differential equations. The practical side of dynamical systems is that they can sometimes be used to describe the behavior of real-world phenomena such as planetary motion, fluid flow, the diffusion of drugs, the behavior of interindustry relationships, and the vibration of airplane wings.

Among my favorite dynamical systems are ones with time-discrete phase planes associated with the cyclic systems: $\dot{x}(t) = f(y(t))$, $\dot{y}(t) = f(x(t))$. The phase-plane diagram for this system indicates the trajectories of points through time. The dot above the x and y indicates a derivative with respect to time. In "phase space," each dimension can represent one of the variables in the differential equation. Try to make pictures where the trajectories of variables (x,y) are plotted to reveal complicated motions. The discretization of these equations for implementation on a computer takes the following simple form (known as the forward Euler approximation): $x_{i+1} - x_i = -hf(y_i)$, $y_{t+1} - y_t = hf(x_i)$. Here, $h > 0$ is a constant known as the *step size* of the numerical solution. I usually keep h small ($h \sim 0.1$). You can try different functions for f; for example, I often use $f(x) = \sin[x + \sin(3x)]$. By plotting the values of x and y as the equations are iterated, you can reveal a multitude of beautiful, chaotic behaviors.

If we let our imaginations freely roam, then we can consider an "attractive fixed point" of this system as an analog of a black hole. An attractive fixed point "sucks" in various surrounding initial points. All points attracted to the fixed point are said to be in the fixed point's basin of attraction. In a sense, the edge of a basin of attraction corresponds to the event horizon. For mathematical systems, the event horizon can have an irregular, fractal shape. I discuss these particular equations in detail in my book, *Computers, Pattern, Chaos and Beauty*.

CHAPTER 15

The star has to go on radiating and radiating, and contracting and contracting, until, I suppose, it gets down to a few kilometer radius, when gravity becomes strong enough to hold in the radiation, and the star can at last find peace.

— S. Chandrasekhar

Black holes are places where God is dividing by zero.

— Anonymous

Black Holes Evaporate

"Sir, you're killing them! The water's boiling hot!"

"Mr. Plex, please don't get yourself so excited. These are thermal bacteria from Venus. They like it hot." You and Mr. Plex stare into the steaming caldron of jumentous liquid. Little black dots undulate and twist on the roiling surface.

A few feet away, Mrs. Plex is sitting on the bed in your cabin, watching as the mist spreads across the room. Bookshelves rise to the ceiling. To your left is a great dark mirror hung over a marble fireplace. Mr. Plex's diamond body sparkles in the reflected light from the mirror.

"Mr. Plex, did you know that black holes evaporate?"

Mr. Plex's brachiocephalic head swivels in your direction. "Evaporate?"

"A black hole behaves as though its horizon has a temperature, and that temperature is inversely proportional to the hole's mass:

$$T = \frac{6 \times 10^{-8}}{M}$$

Here M is expressed in units of solar masses (2×10^{33} grams). The temperature is in degrees Kelvin, that is, degrees centigrade above absolute zero."

Small beads of steam condense on Mr. Plex's diamond body. "Sir, this means that a hole recently formed by the gravitational collapse of a star (which has to have a mass larger than about 2 suns) has a temperature less than 3×10^{-8} degrees above absolute zero." He pauses. "That's pretty damn cold."

"Mr. Plex, watch your language. There's a lady present." You look toward Mrs. Plex, who assumes a reclining position on your bed. She yawns. Her lips are brazenly rouged, and her enormous hazel eyes flash like gems.

You turn back to Mr. Plex. "A mass with a finite temperature radiates energy. Anything that radiates energy is also losing mass. Looking at the equation, you can see that as the black hole loses mass, the emission of energy from the black hole increases and its temperature increases, and thus the rate of mass loss increases. As the mass of the black hole gets small, we have an unstable 'runaway' effect."

Mr. Plex's forearms quiver. "The black hole gets hotter and hotter—"

You wipe away some steam from your forehead. "Which causes M to decrease rapidly—"

"Thus making the black hole even hotter."

Mrs. Plex smiles delightfully as she kicks off her shoes. "When the hole is reduced to a fraction of the size of an atomic nucleus it will be trillions of degrees. The hole will burn up and eventually disappear!"

"Yes, but not in your lifetime, Mrs. Plex. The lifetime of a black hole of mass M is given by:

$$\tau \sim 10^{66} M^3 \text{years}$$

where M is in units of solar masses."

Mr. Plex's diamond teeth are barely visible through the fog. "Sir, this means the lifetime of a black hole is greater than the age of a universe."

"True. However, tiny primordial black holes, about 5^{14}grams(500 million tons) would be evaporating and blowing up even as we speak. We think that some of these tiny black holes were formed when the universe was born in the big bang. They weigh less than 50 billion kilograms, the weight of a small mountain." You pause. "They'd have horizons about the size of an atomic nucleus. Perhaps somewhere in the universe the tiny black holes are exploding."[1]

Mr. Plex gestures to the pot of boiling bacteria. "Sir, can you shut that thing off?"

"Just a little longer, Mr. Plex." You turn to Mrs. Plex, admiring her peach-colored silk and her clattering jewelry.

She rises from the bed and walks toward you, carrying with her a cloud of sweet perfume. You feel a rush of adrenaline, a zeal to discover what she has to say. "Sir," she says, "what would be left over after a hole evaporates?" She says the word *sir* with an ambiguous tone, perhaps humor, but hopefully not condescension. "Could the hole disappear completely, leaving nothing behind, not even the singularity?"

"Mrs. Plex, scientists are not sure."

Mr. Plex's abdomen curls into a tight spiral. "Sir, can we simulate this on a computer?"

"I thought you'd never ask." You reach under your bed and pull out a notebook computer. It feels wet. You wipe off a piece of mucilagenous pizza cheese stuck to the computer's display, and then you toss the computer to Mr. Plex. "Mr. Plex, the decrease in mass of a black hole as a function of time due to thermal emission is:

$$M_t = \left[M_0{}^3 - 3Kt \right]^{1/3}$$

M_0 is the initial mass of the hole. You can also compute the temperature increase as a function of time by using the formula I gave you previously. K is a constant that we can set to 1 to get a feel for how the equation behaves."

Mr. Plex begins to type furiously on the keyboard and instructs the computer to draw a graph of mass and temperature as a function of time for a black hole of 100 solar masses. He also has the computer print out the last few numbers before the temperature values skyrocket beyond the range of the computer. He hands you a printout:

```
Time            Mass      Temperature
333333.29       0.50      1.97
333333.30       0.46      2.15
DANGER!  BLACK HOLE ABOUT TO EXPLODE!
333333.31       0.41      2.42
333333.32       0.34      2.92
DANGER!  BLACK HOLE ABOUT TO EXPLODE!
333333.33       0.21      4.64
```

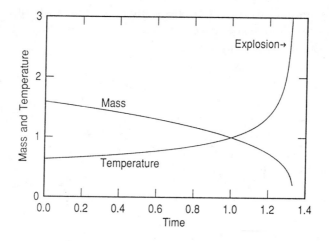

A peek at the gruesome death of a 100 solar mass black hole just as
the temperature values skyrocket and the hole has become anorexic.
As a black hole ages, the temperature explodes, and the mass evaporates to zero. To conveniently scale the axes, temperature is simply
$1/M$, and a large constant is subtracted from the time axis so we can
capture the final seconds of the cosmic funeral.

"Very good, Mr. Plex. Notice that there is a critical time at
which point the mass becomes zero,

$$t_{crit} = M_0^3 / 3K,$$

after which the equation is meaningless."

1. Embedding diagram showing gravitational waves produced by a binary system of two black holes. (See chapter 9.)

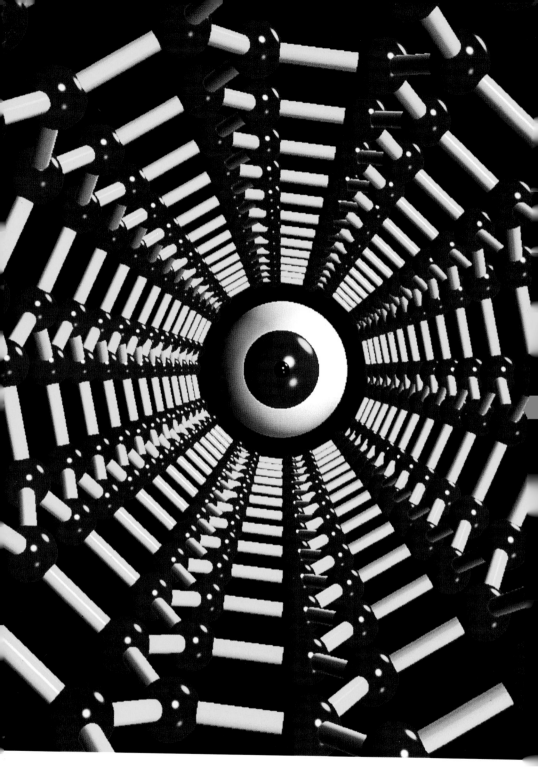

2. Inside the circular Fullerene space tunnel (chapter 4).

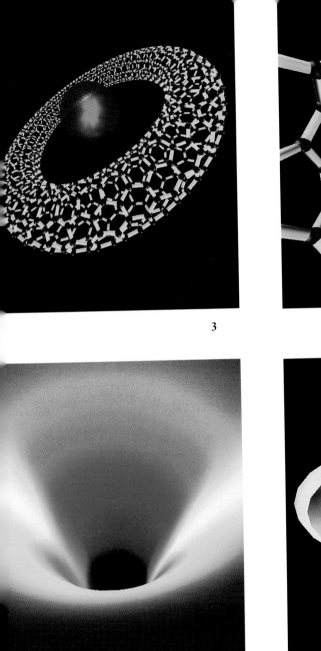

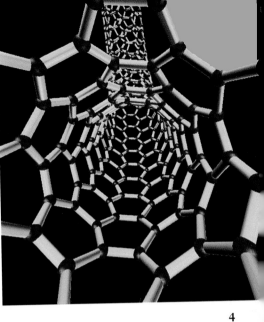

3

4

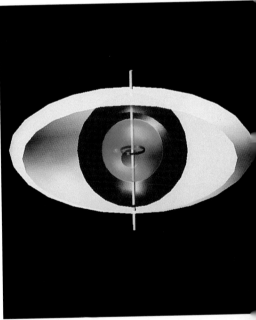

5

6

3. An outside view of the Fullerene space tunnel (chapter 4).

4. Inside a high-altitude Fullerene space tunnel (chapter 4).

5. Embedding representation of a traversable wormhole (chapter 16).

6. Rotating black hole, three-dimensional representation (chapter 7).

7. Three-dimensional embedding diagram for a static black hole (chapter 8).

8. Computer model of a quantum-foam embedding diagram (chapter 12).

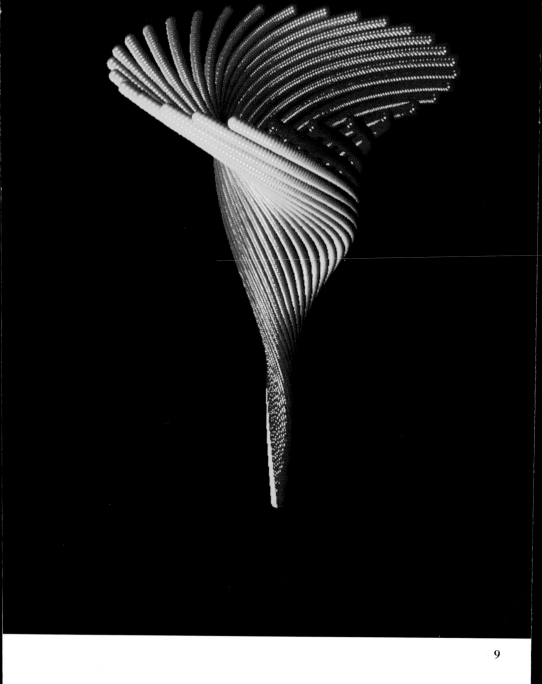

9

9. RLC whirlpool produced by an electrical circuit (chapter 13).

10.   Artistic rendering of the Mandelbrot set (chapter 13).

11. Julia-set "black hole" with fractal "event horizon" (chapter 13).

"This is a wonderful simulation, sir. But you never tell me how you arrive at your equations. You're like a magician pulling them out of a hat. How did you arrive at the formula for mass as a function of time?"

You go to a blackboard on your wall. "Here's a derivation that even your feeble mind should be able to comprehend."

Mrs. Plex grabs a large hunting knife from your bedside drawer. "You will not talk to my husband in such a demeaning fashion." She flashes the knife in front of your eyes. On it are the words "World's Fair 1965."

"Mrs. Plex, I apologize. Just a joke. I wanted to see how closely you were listening to my lecture. I have the highest regard for Mr. Plex."

She puts the knife down. "I can never tell when you're joking. I suppose I overreacted." She pauses. "The heat is stifling. Makes me a little crazy myself."

You turn to the blackboard. "Where was I?"

Mr. Plex comes closer. "Deriving the mass equation, sir."

You nod. "Let's start with the formula for the change in energy of a thermally emitting body through time. You can find the formula in most physics books:

$$\frac{dE}{dt} = -A\sigma T^4 = -4\pi R^2 \sigma T^4.$$

The dE/dt denotes the energy fluctuating through time t. A is the area of the body, and T is the temperature. The minus sign denotes the loss of energy through time. σ is a constant called the Stephan-Boltzman constant, which is equal to $\sigma = 5.67 \times 10^{-8}$ watt/(meter)^2 ($^\circ$Kelvin^4)."

Your voice lingers over the words *Stephan-Boltzman constant*, and this time you are not merely imagining that there is admiration in Mrs. Plex's eyes regarding your intellectual prowess. She appears hypnotized by your eloquence.

Or perhaps the steam is causing her to become disconnected from reality.

You continue. "Now we plug in values for R, T, and E. Here, R is the Schwarzschild radius, $R = 2GM/c^2$. From our previous equation we know that $T = 6 \times 10^{-8}/M$. From Mr. Einstein, we know $E = mc^2$. Combining all of this we get

$$\frac{dMc^2}{dt} = -4\pi(2GM/c^2)^2 \sigma \left[6 \times 10^{-8}/M\right]^4$$

Mrs. Plex's pupils appear to dilate. She is flushed, almost feverish. This realization in turn causes your own heartbeat to increase.

"Collapsing all the constants into one big constant K we get the following:

$$\frac{dM}{dt} = -\frac{K}{M^2}.\text{"}$$

A blush comes to Mrs. Plex's cheeks. You do everything in your power to convince her of your sophistication.

It's time to bring out the heavy mathematical artillery. You integrate the formula:

$$M^2 dM = -K dt$$

to get

$$(1/3) \times (M_0^3 - M_t^3) = K_{t-t_0}$$

She leans closer so that you see the mascara on her long lashes.

$$M^3(t) = M_0^3 - 3K(t - t_0)$$

Mrs. Plex gasps. You feel like running away with her to some remote tropical island. But she is married, and you would never violate the bonds of holy matrimony. Still, the fantasy is an interesting one.

You continue.

$$M_t = \left[M_0^3 - 3K(t-t_0) \right]^{1/3}$$

"If we set $t_0 = 0$, then we finally get

$$M_t = \left[M_0^3 - 3Kt \right]^{1/3},$$

and this is the original equation I showed you, Mr. Plex."

Mr. Plex applauds wildly and picks you up in a fit of emotion. "Very nice, sir! Thank you."

"Uh, Mr. Plex, you can put me down now."

A great cacophony comes from outside. You turn to Mr. and Mrs. Plex. "Must be one of the elephants that escaped from the African habitat. Don't worry about it. Someone will find it."

Mrs. Plex, utterly drenched with sweat, is gazing into your pot of thermal bacteria. "What a nuthouse this place is," she whispers. Then she says in a loud voice, "It's boiling over onto your stove!"

You leap over to the stove and turn it off. "Thanks."

She comes up to you and gazes longingly into your eyes, or so it seems. "Are you doing an experiment with the bacteria's thermal properties? Do the bacteria have a certain degree of intelligence?"

"No, I was just trying to determine the degree to which your husband could withstand an uncomfortable environment." You smile. "He's just so damn polite sometimes. I wanted to see if I could break his elegant facade, rile him up a bit . . ." Your grin widens. "And still he never once complained about the sweltering conditions in my cabin."

Mrs. Plex backs away, screams, and starts looking around the room. "Where did I put that hunting knife?"

THE SCIENCE BEHIND THE SCIENCE FICTION

Black Holes Do Have a Temperature

Black-hole evaporation was first studied by the two Soviet researchers Yakov Zel'dovich and Alex Starobinksy, who, around 1973, discovered that rotating black holes could create particles out of energy and eject them into space. The idea was acceptable because the energy to make the particles could conceivably come from the rotating space bordering the black hole. However, when British theoretical physicist Stephen Hawking tried to develop a proper mathematical treatment of this particle ejection using quantum mechanics, he found, to his surprise, the equations suggested that even nonrotating black holes would emit particles. Because the ejected particles carry energy and this corresponds to temperature, black holes *do* have a temperature.[2]

Mechanism of Evaporation

Quantum physics suggest that short-lived pairs of particles are created in space, and these pairs flicker in and out of existence in very short time scales (10^{-44} second). Such pairs immediately annihilate one another, giving back to space the energy they have temporarily borrowed. The process by which a black hole emits particles involves the creation of virtual particle pairs right on the edge of the black hole's horizon. The black hole's tidal gravity pulls the pair of virtual photons apart, thereby feeding energy into them. One member of the pair is swallowed by the black hole, and the leftover particle scoots away out into the universe.

Siberian X-File

Dr. Robert Wald notes that if a very low mass primordial black hole were to strike the Earth, it would pass right through, absorbing very little matter. However, its strong gravitational field would cause a shock wave through the atmosphere. Some have speculated that a small black hole was responsible for the 1908 Tunguska "meteor" event in Siberia. (Trees were knocked down over a small area covering many square miles, but no meteor crater was found.)

Today experts believe that the black hole explanation is very unlikely because primordial black holes are probably quite rare. Many astrophysicists think that the Tunguska object exploded in midair before striking the ground, and it was probably a small (80-meter diameter) block of ice and stone, similar in composition to the nucleus of a comet. The explosion may have occurred because the material decelerated in the lower, denser atmosphere, thereby converting the block's kinetic energy to heat, vaporizing the ice, and causing the explosion. This is similar to what happened when fragments of Shoemaker-Levy 9 hit Jupiter.

Six Large Power Stations

Tiny black holes with the mass of Mount Everest would have a Schwarzschild radius of only about 10^{-13} centimeters, roughly the size of an atomic nucleus. Dr. John Gribbin notes, "It would be hard for such a black hole to swallow anything at all—and even 'eating' a proton or neutron would be quite a mouthful for it." Nevertheless, as Gribbin observes, the black hole would have a temperature of 120 billion degrees and be radiating away energy at a rate of 6,000 megawatts, equivalent to the output of six large power stations.

CHAPTER 16

There is no question that there is an unseen world. The problem is how far is it from midtown and how late is it open?

— Woody Allen

The thing's hollow—it goes on forever—and—oh my God!—it's full of stars.

— Arthur C. Clarke, *2001: A Space Odyssey*

Wormholes, Cosmological Doughnuts, and Parallel Universes

"Sir, what you propose is madness!"

"This is the day I've been waiting for my entire life, Mr. and Mrs. Plex."

Mrs. Plex comes closer in a tightly fitted and tapering navy blue peacoat. Even in her profile, you can see her deep-set, luminous eyes. "You're *not* going to try to travel to the singularity at the center of a black hole?"

Standing tall on the bridge of your intergalactic zoo, you look from Mr. Plex to Mrs. Plex. It is close to midnight, and most of the animals have gone to sleep, except for the nocturnal species: the bats, the owls, the deep-sea fish.

"Mr. Plex, I've always wanted to enter a black hole or a wormhole and travel to another universe."

Mr. Plex's diamond forelimb taps repeatedly on the floor. "Sir, we've talked about how the embedding diagram of a static black hole connects our universe with a second, parallel universe. You called it an Einstein-Rosen bridge."

You nod. "You have an excellent memory, Mr. Plex. Such a wormhole could also be connecting different areas of our same universe." You pause. "I could survive if the black hole were large enough.[1] The tidal force is actually less for larger holes. Right at the event horizon of a black hole, the relative acceleration between my head and feet would be inversely proportional to the mass of the black hole."

"But, sir, the tidal gravity will still crush you as you get near the black hole's throat." Mr. Plex pauses. "I also hear rumors that the hole between universes is very unstable. Once it forms, it expands and contracts before anything can cross it."

"That's why I've been considering diving into a rotating black hole. Instead of the point singularity, that a static black hole has at its center, a rotating black hole has a ring singularity, a cosmological doughnut of sorts. I could go through the ring without encountering infinitely curved space-time!"[2] You pause. "After going through the ring, I'd emerge into another region of space-time, usually interpreted as another universe. It's even possible to adjust my trajectory through the tunnel to emerge whenever I wish, even thousands of years from now."

Mrs. Plex gazes into your eyes. "I sense a hesitation in your voice."

"Mrs. Plex, some scientists believe the ring singularity spews an intense flux of high-energy particles into the tunnel between universes. Not only might they kill me, but also they might seal off the tunnel."

Mrs. Plex comes a bit closer and puts her hands on her hips. "Then what do you propose?"

"I once gave you a lesson about quantum foam. In the foam, adjacent regions of space are continually stealing and giving back energy from one to another. These cause fluctuations in the curvature of space creating microscopic wormholes."

The tapping of Mr. Plex's forelimb becomes more incessant. "But, sir, those wormholes are too tiny for you to use."

You bring out a bright, metallic object from a locked cabinet. "Yes, but we'll enlarge them with my new invention. It's a gun that spews out something called exotic matter. Exotic matter will enlarge and hold open a wormhole."

Mrs. Plex looks from the gun to you. "Exotic matter?"

"Yes. It has a negative average energy density. My gun gets it from the vacuum fluctuations of the electromagnetic field." You pause, sensing confusion on their faces. "Again, these fluctuations are the random oscillations of a field caused by a tug-of-war between adjacent regions of space stealing energy from one another."

You stop talking when you notice a large diamond ring on Mrs. Plex's finger. "That's quite a rock you've got there, Mrs. Plex."

She hugs her husband. "Why, thank you. We chiseled it out of his forelimb." Mrs. Plex's voice becomes drenched in the sound of the ship's engines. You are bringing the ship to a stop.

Mr. Plex looks nervous. "Sir, you're not taking the ship into a wormhole?"

"No, Mr. Plex. I'm the only one going on this journey." You feel a sense of sudden elation. Almost there. The sound of the engines grows harsher as you throw them into reverse to slow your descent. "Mr. Plex, the good thing about this wormhole solution to Einstein's field equations is that the hole will always remain open, have very small tidal forces, and have two-way travel with no event horizons."

"Sir, does that mean we might see you again?"

"Perhaps, if I succeed. There's no guarantee where and when I'll emerge. Who knows what the alternate universe may look like?"

Mr. Plex's eyes grow moist. "Sir, may we one last time do a simulation together on the computer?"

"Certainly, Mr. Plex. A traversable wormhole can be described by

$$z(r) = \pm b_0 \ln(r/b_0 + \sqrt{(r/b_0)^2 - 1})$$

The space-time embedding diagram has two surfaces and two asymptotically flat regions. It has no horizons, as did the Schwarzschild wormholes we discussed for black holes. The throat is smallest at a minimum value of r, where $r = b_0$. The two flat regions correspond to two universes."

Mrs. Plex withdraws a notebook computer from beneath a potted philodendron plant on the bridge. She starts to hand it to Mr. Plex, but hesitates. "Let me try programming it," she says as she begins to type. After a few seconds a double trumpet shape emerges on the screen.

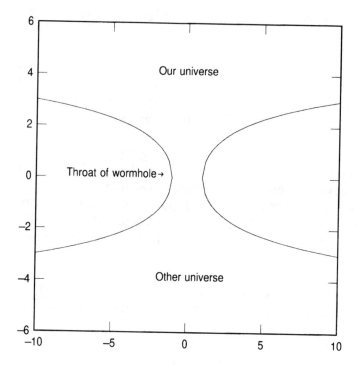

Cross-section of traversable wormhole.

A shiver runs from very base of your spine up to your neck. "Notice that the radial coordinate r has a particular significance. If you place a circle around the wormhole's throat, $2\pi r$ is the circumference. r decreases from $+\infty$ to a minimum value of b_0 as one moves through one universe to the other."

Mrs. Plex takes your hand in hers. Her eyes are bright. "What would the wormhole really look like?"

You give her hand a squeeze. "The embedding diagram is just a convenient way to represent four dimensions on a computer screen. In reality, each mouth of the wormhole would look like a sphere floating in our 3-D universe. Of course, we might not be able to *see* their spherical shapes at all."

The ship has stopped. It is almost quiet. You can hear a pygmy marmoset in an adjacent corridor. You also hear the

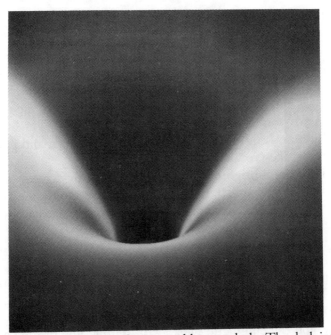

Embedding representation of a traversable wormhole. The dark interior corresponds to the wormhole throat and may lead to another universe.

sounds of Mrs. Plex's breathing. You look from Mrs. Plex to Mr. Plex. "It's time to go," you smile. "Wish me luck."

Mr. and Mrs. Plex help you put on your spacesuit. Mr. Plex hands you the exotic matter gun. Mrs. Plex gives you a hug.

"Sir, your oxygen supply should last for a day."

You nod.

Once out in space you look back at your ship, and see Mr. and Mrs. Plex waving to you from a small portal. You wave back. They look so small. You float away from the ship using small rocket thrusters on your feet, and then you point your exotic matter gun away from your body. A tiny wormhole, about the size of a marble, forms and begins to enlarge. Soon it is larger than your body.

"Here goes," you whisper to yourself. You propel yourself toward the wormhole. "Oh, my God!" you scream.

"The lights!" It is the last thing Mr. and Mrs. Plex hear from you.

You look all around. Gaps start to appear in the heavens, and the stars assemble themselves into gorgeous bands of light. And then the light from the stars changes colors, fractures, replicates. . . . Yellow stars suddenly turn green. The cosmic painting reminds you of something Jackson Pollack would create while high on LSD. And all this lovely play of form and light and color makes you begin to shiver.

Brighter and brighter the lights become. You scream out in ecstasy as the constellations begin to repeat in multiple images. You propel yourself through the wormhole and emerge into a different universe.

But wait! The light is back to normal. Something is odd. Apparently nothing has changed. You turn around and see your intergalactic zoo. The ship is the same. The stars are the same. The more recognizable constellations are unmistakable. Orion's square shoulders and feet, the beautiful zigzagging Cassiopeia, and the enigmatic Pleiades all remind you of life back on Earth where you studied the constellations long ago. You even see Aldebaran, the red star in the constellation Taurus. You remember the time you first saw Aldebaran as a boy growing up in Ajaccio, a town in Corsica, the birthplace of Napoleon.

What should you do next?

You return to the ship and enter its interior. The corridors are unnaturally quiet. Slowly you make your way to your cabin, your footsteps reverberating down the polished halls. You open the door and enter. Everything looks the same. There is your large mirror. Your fireplace. Your books.

You look at the mantel above the fireplace and gasp. There is a photo of you and Mrs. Plex. She is holding your hand. She is wearing a white dress. A wedding dress.

There is a sound from the bedroom. "Oh, honey, I'm glad you're back." Mrs. Plex comes over and gives you a big hug.

"Mrs. Plex, we—We're married?"

"Mrs. Plex?" She wrinkles her brow. "Who is Mrs. Plex?"

You take her hand in yours and note that her diamond ring is of ordinary size and shape. You reach your thumb down the nape of her neck.

She looks into your eyes. "Think we'll ever get off this zoo ship? It's not much fun here."

You hold her hand. "Really? You don't like it here? The work is always interesting. The pay's good. Lots of privacy out here in interstellar space."

"Privacy? You call living in a zoo privacy?" She gestures to the bedroom. You follow.

You throw back your head and scream. How can it be? In your previous universe you were *captain* of your ship.

One wall of the room is made of lucite. People are staring. Above the bed is the sign: "Welcome to the Intergalactic Zoo. *Homo sapiens*, Natural Habitat."

THE SCIENCE BEHIND THE SCIENCE FICTION

In a universe divested of illusions and lights, man feels an alien, a stranger. His exile is without remedy since he is deprived of the memory of a lost home or the hope of a promised land.
— French existentialist Albert Camus

Cosmic Gateways

All of the facts relating to wormholes and black holes in this chapter are derived from recently published scientific papers. The bizarre lighting effects you encountered as you approached the wormhole are based on scientific facts discussed in the "Space Mirror" section.

Much of the recent research on wormholes has been conducted by Kip Thorne, a cosmologist, and Michael Morris, his graduate student. In a scientific paper published in the *American Journal of Physics* (see "Further Reading"), they developed a theoretical scheme for inter- and intra-universal travel. In their research, they elegantly prove the theoretical possibility for wormholes that bridge the universe. These cosmic gateways can be created between regions of the universe trillions of miles apart and would allow nearly instantaneous communication between these regions. Interestingly, Carl Sagan in his novel *Contact* also uses these same kinds of wormholes to traverse the universe. Additionally, the television shows *Star Trek: The Next Generation*, *Star Trek: Voyager*, and *Star Trek: Deep Space Nine* have all used wormholes to travel between faraway regions of space. In *Star Trek: Deep Space Nine*, a space station stands guard at one end of a stable wormhole.

Not only can the Morris-Thorne wormholes be used for space travel, but also, if properly constructed, they may be used for time travel. For details, see the paper by Morris, Thorne, and Yurtsever in the "Technical Articles" section.

Exotic Matter, Throat Tension, and Space Mirrors

To fully realize a wormhole, Morris and Thorne calculated various properties of matter required to form the wormhole's throat. One property of interest to them was the tension (i.e., the breaking strength) of matter needed to keep the wormhole open. What they found was that the required tension would be very large. As Paul Halpern in his book *Cosmic Wormholes* notes, for a throat that is four miles across, the quantity of force needed is 10^{33} pounds per square inch. This would be more than the pressure of a trillion boxes, weighing a trillion tons each, placed in the palm of your hand. Larger wormholes with wider throats would have more reasonable values for throat tension.

Morris and Thorne also found another difficult situation to overcome when considering the matter needed to form the gateway. The tension needed to keep the wormhole open must be 10^{17} times greater than the density of the substance used to build the wormhole. According to current science, there is no matter in the universe today that has breaking tensions so much larger than its density. In fact, if the tension of a piece of matter were to rise above 10^{17} times its own density, physicists feel that material would begin to possess strange attributes, such as negative mass. Because of these unusual characteristics, the type of matter needed to keep wormholes open has been called exotic matter. Matter of this type may exist in the vacuum fluctuations of free space. To make wormhole construction easier, it may be possible to construct the entire wormhole out of normal matter and use exotic matter only in a limited band at the throat.

About a year after Morris and Thorne delved into wormholes, Matt Visser of Washington University developed a wormhole model that "looked" more like a rectangular spool of thread than the hourglass shape of Morris and Thorne. A rectangular hole in the middle of the spool corresponded to the wormhole, the gateway between two regions of space. In this model, described in *Physical Review*, the wormhole's boundaries are straight and can be made as distant from one another as desired. Exotic matter could therefore be placed faraway from the wormhole travelers to minimize risks. This rectangular spool model may be more stable than the Morris-Thorne model, and gravitational tidal forces on passengers would be less of a concern. How would the end of a Visser wormhole appear to you as it floated in space? It would look like a dark, rectangular box. You could approach this black prism and enter it near the center. Almost immediately you would exit a similar dark prison comprising the other side of the wormhole. These two prisms would be connected via the fourth dimension along the shaft of the spool that comprises the wormhole's throat.

The exterior of the Visser wormhole acts like a giant mirror. Light shining upon it would bounce off as though it hit a reflecting material. Visser also proposed another mathematical model for a wormhole that resembles two coreless apples. The inner walls of the fruit are connected along the fourth dimension. You can read more about this structure in Halpern's *Cosmic Wormhole* book, or in Visser's original scientific paper.

Wormhole Light Show

It is as if a large diamond were to be found inside each person. Picture a diamond a foot long. The diamond has a thousand facets, but the facets are covered with dirt and tar. It is the job of the soul to clean each facet until the surface is brilliant and can reflect a rainbow of colors.
—Brian Weiss, M.D., *Many Lives, Many Masters*

The psychedelic light show you experienced when you left the Plexes and approached the wormhole is based on scientific fact. As you get nearer to a wormhole throat, gaps would start to appear in the sky as the stars transform into bands of light. As you go even closer, the bands multiply, spreading across the sky in a series of images. Star systems appear as multiple images. This replication of images is caused by gravitational lensing (see chapter 4). The huge mass of the wormhole causes light rays to focus themselves into clusters. The colors of the stars become blueshifted. Red becomes orange, orange becomes yellow, and so on. Again, the large gravitational force of the wormhole causes the incoming light to increase its frequency. Lights appear fractured, and they brighten as you approach the wormhole.

Animated, Inflating Wormholes Using Mathematica

It is possible to make animated embedding diagrams of traversable wormholes in an inflationary (expanding) universe simply by multiplying the embedding diagrams formulas in this chapter by a time-dependent scale factor e^{χ^t} where t is time and χ is a constant. Physicist Thomas Rowan recently created such an animation using the software Mathematica. For details, see the "Technical Articles" section. The availability of these kinds of programming tools on desktop computers should allow students and researchers to more fully exploit the beautiful geometrical features of wormhole theory.

Can Black Holes Be Used as Interstellar Gates?

In principle, you can enter a rotating black hole and actually steer away from the singularity and out to a new, asymptotically flat region of space-time, although there are several practical problems with this approach, as discussed by you and Mr. Plex. Interestingly, if you were to dive through the ring singularity, gravity would reverse as you passed through the ring, turning into a repulsive force that pushes you instead of pulling. Dr. John Gribbin notes that an astronaut who dived through the ring, but stayed close to it and circled around the center of the black hole in an appropriate orbit, would be traveling backward in time. He says, "The saving grace, from the point of conventional physics, is that even if you did this, then dived back through the ring and on out of the rotating black hole, you still couldn't go back to the same region of space that you started from." Gribbin also notes that if the hole rotated quickly enough, its event horizons would disappear and leave a naked singularity in the form of a ring. In theory, it would be possible not only to travel through the ring but also to *look* through it from far away using a powerful telescope. Theoretically speaking, a rotating black hole may connect our universe to itself at many places and times.

Unfortunately for black-hole travelers, Kerr tunnels formed by the ring singularity in a rotating black hole, although very stable in isolation, would be quite sensitive to outside disturbances, which would lead to structural collapse. Even if you could survive the inner horizon singularity, the presence of an average-sized human would probably collapse these tunnels, immediately squashing the adventuresome traveler. These gravitational forces would be accompanied by lethal bursts of radiation caused by the rapid crushing of the captured material. However, some astrophysicists have speculated that advanced civilizations might develop the technology to prevent the sensitivity of Kerr tunnels to disturbances, thereby preventing the decay of the tunnels.

> It is known that there are an infinite number of worlds, simply because there is an infinite amount of space for them to be in. However, not every one of them is inhabited. Therefore, there must be a finite number of inhabited worlds. Any finite number divided by infinity is as near to nothing as makes no odds, so the average population of all planets in the Universe can be said to be zero. From this it follows that the population of the whole Universe is also zero, and that any people you may meet from time to time are merely the products of a deranged imagination.
> —Douglas Adams, *The Restaurant at the End of the Universe*

P O S T S C R I P T 1

The expression black hole *is often still no more than a sumptuous disguise for our ignorance.*

— Jean-Pierre Luminet, *Black Holes*, 1992

Our entire universe may slowly stop expanding, go into a contracting phase, and finally disappear into a black hole, like an acrobatic elephant jumping into its anus.

— Martin Gardner, "Seven Books on Black Holes," 1981

I'll tell you what the Big Bang was, Lestat. It was when the cells of God began to divide.

— Anne Rice, *Tale of the Body Thief*

Could We Be Living in a Black Hole?

Over the past few decades, various physicists and astronomers have wondered whether we could actually be living in a black hole and not even know it. The answer to this question can be viewed from a variety of perspectives, and I have found that scientists do not agree on a single answer. Some of the following material is based on discussions I have had with physicists and astronomers on the Internet.

As background, first consider that both the motion of galaxies and the detection of uniform cosmic radiation indicate that the universe has been expanding for the past 15 billion years, ever since the big bang. This expansion is particularly evident in the ever-widening distance between clusters of galaxies. To better visualize the expanding universe, you can partially inflate a balloon and tape paper clips to its surface to represent galaxies. As you inflate the balloon, each paper clip "sees" neighbor clips receding from it. In fact, the clips twice as far away recede twice as quickly. Note that the paper clips themselves are not expanding, but rather each one finds itself at the center of an expanding pattern of clips.

In the future, there are two possible fates for our universe. If God is an infinitely patient balloon-blower, the "universal balloon" may expand forever. If God stops and takes a breath, our universe may contract after a finite amount of time and finally coalesce in a sort of inverse big bang known as the "big crunch." In this latter scenario, gravitation acts on the expanding universe, causing it to contract back to a single point about 100 billion years from now.

If our universe is a "closed" universe that eventually recollapses, might the entire universe be a black hole? Robert M. Wald, a professor of physics at the University of Chicago, has speculated on this precise question. In his book *Space, Time and Gravity*, Wald notes that there is a close analogy between the behavior of a body that undergoes complete gravitational collapse and the collapse of the entire universe. In a finite time, both the collapsing body and the universe result in infinite density and space-time curvature. However, a black hole requires the existence of regions of space-time outside the hole to which nothing inside can escape. Otherwise the entire notion of an event horizon becomes trivial.[1] Therefore, Dr. Wald believes the concept of a black hole cannot be meaningfully applied to the universe as a whole.

John Gribbin, who holds a doctorate in astrophysics from the University of Cambridge, approaches the question from a slightly different perspective. As background, consider that the greater the volume of a black hole, the lower the density of matter needed to trap light. In fact, it is not even necessary for black holes to be dense. For example, consider that the density of a sphere is defined as:

$$\frac{M}{(4/3) \times \pi \times r^3}.$$

For a black hole, we can get a qualitative feel for how density behaves by substituting for the radius r the value of the Schwarzschild radius, $2GM/c^2$, where G is Newton's gravitational constant, M is the mass of the black hole, and c is the speed of light. Therefore, the density of a spherical hole is defined by:

$$\frac{3c^6}{32\pi G^3 M^2}.$$

(For a body with a mass equal to the that of the sun, the Schwarzschild radius is about 2 miles.)[2] This suggests that the more massive a hole, the less its density. In fact, a black hole of several billion solar masses would have a density 100 times less than that of water. Therefore, despite commonly held views to the contrary, a black hole does not have to be extremely dense; it simply has to be sufficiently compact to imprison light. Moreover, black holes can be made out of any substance and have any density, so long as there is enough matter to fill an appropriately sized sphere. In his book *Black Holes, White Holes, Wormholes,* Dr. Gribbin notes:

> A star system like our Milky Way Galaxy but larger, containing . . . billions of stars spread over a sphere with a radius of thousands of light-years, could have an overall escape velocity greater than that of light, even though the stars, planets, and people, within the system were in no way unusual. We could be living inside a black hole and not even notice.

Kip Thorne, a professor of theoretical physics at the California Institute of Technology, wrote to me personally when I asked him: Can we be living in a black hole?

> We know approximately how much mass there is out to cosmological distances, and nowhere short of cosmological distances (10 billion light years) does the mass add up to enough to produce an escape velocity greater than light.
> At about 10 billion light years the amount of mass could be large enough; we aren't sure. However, the region inside that distance has galaxies expanding away from each other, which is not what would happen inside a black hole. Instead, inside a black hole the galaxies would have to be moving toward each other. Thus we could, in principle, be inside a "time-reversed" black hole, i.e., a white hole; but that seems rather improbable.

Let me give another perspective on the question. Jean-Pierre Luminet (a French specialist in black holes, astronomer at the Meudon Observatory, and author of *Black Holes*) believes that in about 10^{27} years all the extinguished stars will have coalesced in the center of the galaxies and formed massive black holes of 10^{11} solar masses (i.e., they will be 10^{11} times as massive as our sun). Much later, in 10^{31} years, galaxies will have dispersed their orbital energy by gravitational radiation and fused to form supergalactic black holes of 10^{15} solar masses.[3]

Luminet notes that if the universe is to be a closed system, its minimum density corresponds to a 10^{23} solar masses black hole with a radius of 40 billion light years. In the observable universe, the largest distance traveled by light does not exceed 15 billion light years. Luminet's conclusion is that the observable universe is inside its

Schwarzschild radius. If the universe is closed, Luminet suggests that there should be an outside world within which our universe is a region hidden inside a black hole.

If our universe started as a black hole, how did it become a black hole in the first place? Was it formed by the gravitational collapse of a 10^{23} solar masses "superstar"? In this case, Luminet suggests that the exterior cosmos would not be empty, and entire galaxies could fall into our universe. On the other hand, perhaps our initial tiny universe was a minuscule primordial black hole in an exterior universe. (As discussed in other chapters, some researchers believe these subatomic primordial holes may exist scattered throughout our current universe.)

What can we know about life inside a black hole? The laws of general relativity predict that all the mass within a black hole converge to a central singularity. However, these laws are generally considered incomplete, and because physicists know even less about the laws of quantum gravity, some researchers feel we know little about the laws governing matter in a black hole when very close to the singularity. Luminet suggests the possibility that the expanding-contracting black hole universe implies that gravitational collapse can somehow be halted inside a black hole, before the mass converges to the singularity. More fascinating is the idea that black holes call into question the uniqueness of our universe. Perhaps there is a hierarchy of universes, black holes within black holes, in a mind-numbing progression of worlds within worlds. Recent theories in physics allow for the existence of such "bubble universes."

Astrophysicst William J. Kaufmann, III, author of *Black Holes and Warped Spacetime*, believes that if the universe is closed, then we are living inside a black hole. He states, "The history of the universe simply consists of erupting from a *past singularity* 20 billion years ago, only to plunge into a *future singularity* 130 billion years from now." Everything will eventually be crushed out of existence at the final cosmic singularity. Perhaps those of you with religious inclinations will wonder whether God rules the expansion phase of the universe, while the devil rules the destruction phase.

Dr. Kaufmann does agree that the hypothetical cosmic black hole in which the universe swallows itself is fundamentally different from any of the usual black holes created by stellar implosion. With ordinary black holes, there is plenty of flat space-time far from the hole. As the collapsing universe swallows itself, there is no asymptotically flat space-time to which the cosmic black hole is connected.

Note that nothing from our universe will survive this big crunch. Possibly, the most basic physical constants and quantities will be forever lost. What will be the value for the speed of light or the gravitational constant in the next universe? Dr. Kaufmann speculates that the next universe, "born from the condemned ashes of our cosmos," would be totally unrecognizable to us.

Several scientists on the Internet have informally conversed with me regarding the central question of this chapter, and I think you will find their ideas illuminating as they reiterate the central ideas already expressed in this chapter. Bob Loewenstein from the University of Chicago has written to me:

> A black hole is defined as something from which light cannot escape. Usually this definition means that there is sufficient mass within a certain radius so that the escape velocity at that radius is greater than the speed of light. An easy formula for the relationship between this Schwarzschild radius and the mass within the radius is $r = 2GM/c^2$. If the "radius" of the galaxy is roughly 40,000 light years, and if the mass contained within that radius is greater than approximately 10^{17} solar masses, then indeed the Milky Way is a black hole. This is rather unlikely, since the mass is closer to 10^{11} or so.

Brian Crabtree from the Netherlands says:

> In his book *Black Holes, White Holes, Wormholes*, Gribbin is making the point that the only condition necessary to establish a black hole is that the escape velocity of an interior object exceed the velocity of light. Therefore, a super-huge galaxy or group of galaxies might meet those conditions; however, when viewed remotely, the set would appear like any other black hole except, of course, it would be unusually massive.
>
> We can only speculate about conditions within the black hole because it is impossible to transfer any information from inside across the boundary. If our super-massive galaxy, discussed above, formed a black hole, then it would be cut off from the rest of the universe and in fact

form a "new universe." If you are within the event horizon when a black hole forms, there is no requirement for the laws of physics to break down as far as you are concerned—except that you are in close proximity to a very large mass that has just left your previous universe.[4]

Finally, Evans M. Harrell, a mathematical physicist from the Georgia Institute of Technology, wrote to me:

Gribbin suggests that we could already be within the capture basin of a black hole. In other words, we may all eventually end up crammed into the singularity. If the black hole is large enough, it is possible that we would have no clue of our fate—presumably billions of years hence.

As suggested in earlier chapters, in the late stages of gravitational collapse, all the mass of a black hole is concentrated at its singularity. A black hole's singularity, roughly 10^{-33} centimeters in size, is surrounded by pure emptiness, aside from tenuous interstellar gas that is falling inward and radiation the gas emits. There should be a void from the singularity out to its event horizon.[5]

Giant Black Holes

Even if it is unlikely that our universe is a giant black hole (see note 5), the theory of gravitational collapse allows us to imagine black holes of a billion solar masses and larger. How could such behemoths be formed? Certainly not from the gravitational collapse of a star, which could produce a maximum mass of about 10 times that of our sun. One possibility is that a giant black hole is created by the condensation of lumps in the early universe. It's also possible that a 10-solar-mass black hole acts as a seed for a larger hole by growing as surrounding matter is sucked in. It's even possible that giant black holes are produced by the collapse of multiple stars in a star cluster. One likely place for a giant black hole is at the center of a galaxy where matter is more concentrated than in other regions of the universe. Another possible site is in the center of globular clusters—dense collections of several hundred thousand stars within a diameter of less than 150 light years. At the center of these luminous balls of stars, the star density is 20,000 times greater than it is in our own vicinity of the galaxy. Black holes of thousands of solar masses could form in the center of globular clusters as surrounding stars are sucked in.

Sometimes when I gaze up at the night sky, I try to look for the constellation of Hercules, which has a famous globular cluster. I wonder what it would be like to live on a planet orbiting a star in a globular cluster. Would there be no dark night sky? Would astronomers know far less about the universe because distant stars could not be seen through the surrounding "pollution" of light? What would it be like to live right on the edge of a cluster's central black hole? Perhaps inhabitants would gaze up at the sky and see the black hole's white-hot glow, from super-heated gas, reddened by intervening dust. If scientists on such a planet realized their fate, how would this affect politics and religion? If we had lived on such a star, what strange mythologies would have evolved? Some of the visual aspects of life in a globular cluster can be explored using computer graphical simulations discussed in my book *Chaos in Wonderland.*

Evidence for giant black holes continues to increase. Toward the end of 1995, astronomers used the Hubble Space Telescope to confirm the presence of another super-massive black hole in the universe. The black hole, and an 800 light-year-wide spiral-shaped disk of dust fueling it, are slightly offset from the center of their host galaxy, NGC 4261, located 100 million light years away in the direction of the constellation Virgo. It's possible that material from the off-center disk, falling onto the black hole, propelled the hole away from the galaxy's center.

Astronomers have suspected for at least a decade—ever since radio telescopes detected twin, oppositely directed jets of radiowaves streaming out of the galaxy's center—that NGC 4261 harbors a black hole. By measuring the speed of gas swirling around the black hole, the team of astronomers was able to calculate its mass to be 1.2 billion times the mass of our sun, yet concentrated into a region of space not much larger than our solar system.

P O S T S C R I P T 2

Black holes are the bane of modern astronomy. Invisible by definition, their existence has proved difficult to substantiate, yet black holes have captured the popular imagination in a way that no other astronomical objects have succeeded in doing. Black holes are time machines, as well as openings to other universes.

— Joseph Silk, University of California at Berkeley

In some sense, black holes are the best of all worlds, subdued enough so that we can attempt to understand them and bizarre enough to still fire our imaginations with infinite possibility.

— Warren G. Anderson, University of Alberta

The Grand Internet Black-Hole Survey

Black holes continue to fascinate both laypeople and seasoned astronomers. In this chapter, I quote astrophysicists from around the world who have responded to some of my specific questions regarding black holes, and I thank them for permission to reproduce excerpts from their comments. My questions were sent through electronic mail and often posted to electronic bulletin boards. Two common sources for such information exchange were "sci.astro" and "sci.physics"—electronic bulletin boards, or "newsgroups," that are part of a large, worldwide network of interconnected computers called Usenet. The computer users exchange news articles with each other on a voluntary basis.

The five questions I asked were:

1. What is the most common misconception the public has about black holes?
2. Why do you feel that the public is fascinated with black holes?
3. Will a black hole ever serve a useful purpose for humans or some other advanced civilization, or is this purely science fiction?
4. What unsolved problem would you like to have solved in your lifetime regarding black holes?
5. What aspects of black holes do you personally find the most fascinating?

Steve Crisp teaches introductory and advanced observational astronomy at North Carolina State University. His training and research is in the history and rhetoric of science and technology. Here are his views regarding the five questions:

1. *Misconceptions:* The two most common misconceptions are that black holes have volume and that they are very hot. As any self-respecting astrophysicist will tell you, black holes are singularities (or mathematical points) and their temperature hovers at absolute zero.
2. *Public fascination:* Fascination with the unknown.
3. *Useful function:* A black hole might be useful if it were to venture close to the sun and establish a stable orbit, and also if we were able to develop a technique of harnessing gravitational energy. However, the chance of this happening within the lifetime of our species is extremely slim.
4. *Unsolved problem:* I would love to see empirical proof supporting or refuting Hawking's hypothesis for the evaporating black hole.
5. *Personal fascination:* It is fascinating that the sheer mass of tens of thousands of suns is capable of existing as a subatomic singularity.

The following is from a researcher at the Mullard Radio Astronomy Observatory in Cambridge who wishes to remain anonymous:

1. *Misconceptions:* The greatest misconception is that they're totally black (i.e., invisible).
2. *Public fascination:* I'm not sure that the public *is* fascinated with black holes. I would say they are *interested* because black holes sound like science fiction but real scientists seem to take them seriously. Possibly the public is also interested because the basic idea of a black hole is a fairly simple one and easier to grasp than a lot of supposedly cutting-edge physics. Some people find exciting the idea of an inevitable terror lurking invisibly in space.
3. *Useful function:* Nobody's going to be using any preexisting massive black holes as energy sources in the near future. A carefully managed black hole might be an energy source for an advanced civilization that had complete control over its environment and was ubiquitous throughout a galaxy. The formula $E = mc^2$ determines the limit of this energy, and a rotating black hole can only give us about 10 percent of this.

4. *Unsolved problem:* We do not really know how (and if) black holes produce the immensely powerful beams of plasma that make radio-loud quasars and radio galaxies.

5. *Personal fascination:* See question 4, the subject of my research.

William S. Lawson is a physicist trained in plasma physics but currently working on semiconductor simulation at Lawrence Livermore Lab. His interest in gravitation dates back to the time he enrolled in Kip Thorne's class in 1979–80, during his senior year at Caltech. He says:

1. *Misconceptions:* There are two common misconceptions among science fiction writers. One is that matter can go into a black hole and stay there indefinitely, being retrieved at some later time. The other misconception arises from the fact that an object appears to take an infinite amount of time to fall through the event horizon when viewed from a distance. Some science fiction writers have taken this to mean that it is possible to drop an object onto a black hole and retrieve it at one's leisure. In fact, the object rapidly reaches a point of no return where one cannot catch up with it before it crosses the event horizon.

2. *Public fascination:* Black holes are exotic. Things go in but don't come out. They warp space and time. They are the corpses left from the most violent events known, and produce some violent effects themselves.

3. *Useful function:* It is unlikely that we will ever be able to get up close and personal with a black hole, but black holes may be immensely useful to astronomers in some kind of measurement capacity—from our safe distance.

4. *Unsolved problem:* The biggest unresolved problem is whether black holes exist, and are they accurately described by Einstein's theory? A yes to both questions would be an immense step forward for physics, since a black hole is one of relativity's most extreme predictions.

5. *Personal fascination:* Black holes challenge, in a concrete way, one's notions of time. The minor discrepancies in the rates of passage

of time that one finds in the ordinary parts of the universe can be imagined as perturbations of some absolute true time. This is not true for the *extreme* warping of space-time around a black hole.

The following is from Mark Higgins, a Ph.D. student in astrophysics at the Queen's University Astrophysics Department in Kingston, Ontario, Canada. He says:

1. *Misconceptions:* The biggest misconception is that black holes are just big cosmic vacuums, and if we're not careful we might get gobbled up. When the evidence for a black hole at the center of galaxy M31 was just released, a lot of people asked me whether we were in serious danger here on Earth.
2. *Public fascination:* I think people like black holes because they typify science fiction and fantasy.
3. *Useful function:* Some sorts of black holes *might* be able to be used as time machines; however, there are many theoretical problems to deal with here (not to mention practical problems like being ripped into your component subatomic particles).
4. *Unsolved problem:* Many scientists are uncomfortable with the idea of a black hole's singularity found at the center where the laws of physics break down. I'd like to see someone solve this problem. (The solution will most likely come from some a unification of quantum mechanics with general relativity.)
5. *Personal fascination:* Space-time gets curved so strongly in a black hole that quantum effects are bound to become significant, and this is where it's important to figure out some sort of grand unified theory. How this will happen is what really intrigues me.

From Warren G. Anderson, a doctoral student doing research into quantum aspects of black hole space-time in the Physics Department at the University of Alberta:

1. *Misconceptions:* Almost certainly the largest misconception is that black holes exist. Now let me set the record straight. I don't mean that stars do not collapse and form trapped surfaces, but rather that the black holes that we physicists study are only approximations of these collapsed stars at late times. In the strictest sense of

the word, there are no black holes. However, note that what we are learning from our approximations should still describe what's really out there very well. On the other hand, when we say that black holes have this property or that property, often we are describing the properties of our idealized models and not about the properties of collapsed stars. I think physicists are largely to blame for this, since we talk about these properties from our models as though they really exist in all generality, rather than as approximations. For example, we talk about evaporating black holes, when, in fact, once evaporation is included, the properties that make a model a black hole are lost. An evaporating black hole is really an oxymoron.

2. *Public fascination:* Science has been used for centuries as a tool to explain our experiences in a coherent and logical way in order to demystify our world. However, over the past century, science, and in particular physics, has stretched into realms where we cannot have common experiences. What we have found is that the universe is far more interesting and "mystical" than we ever could have imagined. I think that phenomena from these realms, such as black holes, satisfy our need for the mystical and fantastic while at the same time allow us to believe we are understanding our universe in a coherent and logical way. In some sense, black holes are the best of all worlds, subdued enough so that we can attempt to understand them and bizarre enough to still fire our imaginations with infinite possibility.

3. *Useful function:* I think that some day a black hole—a physical one, not a mathematical model—could serve as a fundamental laboratory where we test our theories. Whether they will have more mundane and practical applications is hard to say. However, the time scales we are discussing—at least hundreds of years—are so large that any discussion of how they will be useful might as well be science fiction.

4. *Unsolved problem:* I would like to know the quantum state of a black hole. This will entail the development of a quantum theory of gravity, which we have not yet succeeded in formulating, despite over 30 years of constant work. It may be a bit much to ask for in my lifetime, but I'm hopeful.

5. *Personal fascination:* From our crude marriages of quantum theory and black holes we have already learned that there is an intimate relationship between black holes and the basic principles of thermodynamics. We talk these days about the entropy, specific heat, and temperature of black holes. This is a clue that there are very deep and subtle relationships between these aspects of nature that have not been discovered before. This type of unification in physics has been the "Holy Grail" of physics for most of this century, and black holes are proving to be the best source of information for this unification. Right now, the major drawback is that the marriages between quantum theory and black holes are very flimsy and makeshift, but if and when we properly understand how thermodynamics and black holes fit together, I believe we will find important clues as to how very fundamental aspects of our universe are linked.

From Erik Max Francis, a freelance science fiction writer who is also involved in quality assurance at Adobe Systems, Inc.:

1. *Misconceptions:* The most common misconception is that a black hole somehow magically "sucks" things into it. A black hole attracts other bodies just like all masses, but the Earth doesn't spiral into the sun merely because the sun attracts it. The Earth is in a stable orbit. The same is true of particles in orbit around a black hole; they're in orbit, just like the Earth is in orbit around the sun.

   The reason that objects *do* end up spiraling into a black hole is not only because of gravity, but also because of friction. Generally there's lots of material in orbit around a black hole at any time. This material rubs against other material and creates heat. The heat comes from the kinetic energy of particles involved, and naturally those particles must slow down. A slower speed means a lower orbit, so, over time, particles in an accretion disc tend to get swallowed up by the hole. But there's nothing special about the gravitational effects of a black hole that cause the particles to be fall into it.

2. *Public fascination:* The public is excited about black holes because the holes are contrary to everyday experience. The reason special relativity, general relativity, and quantum mechanics get so much

interest is that not only are they so counterintuitive and contrary to human experience, but they also are three of the most successful theories in history. Black holes are a result of general relativity, and black holes are definitely strange—but more than being just interesting curiosities, it seems that they might well exist in the universe, and they are useful for explaining a variety of phenomena.

3. *Useful function:* It is, of course, impossible to speculate into the far future, but I am aware of only two ways of extracting energy from a black hole. One way to extract energy is via the Penrose process, which requires that the black hole be rotating. The extracted energy comes from the rotational energy of the hole. The other method is via Hawking radiation, where very small black holes (much smaller than the black holes created by stellar collapse) emit large amounts of radiation due to a combination of quantum mechanics and general relativity.

4. *Unsolved problem:* The exact nature of the central singularity is of interest. Is there really a singularity (a point of infinite spacetime curvature and density), or does quantum mechanics somehow intervene and prevent the singularity from being created, resulting in some other less extreme form of gravitational curvature? To answer this question we must have a solid theory of quantum gravity, which at present we do not yet have.

5. *Personal fascination:* I am fascinated by the way relativity applies to black holes. If you were to dive into a black hole, you would pass through the event horizon in finite proper time (that is, time as measured by your watch, or by your internal sense of time). In fact, you would hit the singularity in a finite amount of proper time, although you will die quite horribly due to the gravitational tides before you get near the singularity, at least for a black hole formed from the endpoint of stellar evolution.

On the other hand, from the point of view of someone very far way from the black hole—that is, outside the black hole's gravitational influence in what's called "asymptotically flat spacetime"—a second observer would watch you fall toward the black hole and begin to slow down as you approached the event horizon. In fact, from his point of view, you would never even reach the horizon, and very quickly your image would be redshifted and dimmed so that you could not be detectable.

We can explain this apparent contradiction. The distant observer is perceiving your image, not you. All images are carried by light. Therefore, a remote observer detects you by the light coming or being reflected from you. A black hole can be considered as distorting spacetime in such a way that space-time is flowing into the black hole. It is easy to see why the event horizon is a point of no return in this model. The event horizon is the radius around a black hole where "space-time is flowing inward at the speed of light," so to speak. Since nothing can travel faster than light relative to the space-time it's located in, this means that nothing can ever exit the black hole once it has fallen within the event horizon.

Now we can explain what is happening with your image as you approach a black hole. Consider the case where you are falling into a black hole, and emitting light via a flashlight. As you get closer and closer to the event horizon, the light that you are emitting (or reflecting) to an outside observer is taking longer and longer to get out of the black hole. Since light moves at exactly c, and the closer you get the horizon the faster space-time is "flowing inward," the relative propagation of light out of the hole is slower. At the event horizon itself, any photon emitted out of the black hole will just sit there, since it's moving outward at c, but space-time is "flowing inward" at c. The end result is that your image appears to forever hover just outside the horizon, even though you've long since fallen in and met your doom at the singularity.

Marc Hairston is a space physicist in the Center for Space Sciences at the University of Texas at Dallas, and he teaches an introductory astronomy course. He finds that his students are fascinated with black holes as they start the course.

1. *Misconceptions:* Almost all of my students are convinced that black holes are cosmic vacuum cleaners that reach out across light years to suck in material. When I begin the black hole lecture, we have already discussed the Newtonian idea treating the mass of the sun

as if it were a point mass located at the center of the solar system, so I ask them to do a thought experiment. What happens to the Earth's orbit if all the mass of the sun were to shrink to a point and become a black hole? Inevitably everyone says the Earth would be pulled out of its orbit and sucked into the hole. When I tell them that *nothing* would happen to the orbit (and then reiterate the Newtonian idea of considering all mass concentrated at the center of objects), I have actually received open-mouthed stares from some students. "But it *has* to have a more powerful gravity," some of them wail. "It's a *black hole*." The most difficult concept for the students to understand about black holes is that the "region of danger" around a black hole is relatively small.

2. *Public fascination:* The fact that hard science predicts the existence of some region of space into which things go, but never come back out, seems magical to the public. The sad thing is there is so little else in science that the public thinks as "wondrous" or "magical." (It is our own fault for not communicating the wonder of science to the public and to the students in our schools . . .)

3. *Useful function:* If you accept Kip Thorne's ideas of building a stable wormhole with special devices to keep both ends open, then I guess you could have something as seen in the TV show *Deep Space Nine* where wormholes are used as tunnels to other parts of the universe. But frankly I doubt we will ever have the technology to do such a thing, even if the laws of physics say it's possible.

4. *Unsolved problem:* What happens inside the event horizon is unknown. Obviously, even if we could study a nearby black hole, we'd never be able to find out what happened to a probe that went inside. Therefore, we're limited to the theoretical models, but they keep changing, so my question is, "Will the models ever converge, or will we be forever in an undecided state about what *really* happens inside the event horizon?"

From Daniel Fischer, an astronomy writer, editor of the German astronomy journal *Sterne und Weltraum*, publisher of the weekly newsletter *Skyweek*, a student of Professor W. Kund (the famous Bonn astrophysicist who has developed alternative models for galactic centers and X-ray binaries):

1. *Misconceptions:* That black holes are the best explanation for the activity in galaxies and, in particular, jets.
2. *Useful function:* Black holes already serve a very important purpose for astronomers who need money! Consider yourself an astronomer. Want to have your observing proposal accepted, or a new expensive telescope? Want to justify spending millions of tax money? Just include the phrase, ". . . and I will find black holes . . .," and it will work. Don't believe it? Just look, for example, at the press releases from STScI . . .
3. *Unsolved problem:* The biggest unsolved problem is whether there are *any* black holes in the universe. In my opinion, the evidence so far is slim, and even black-hole aficionados admit this fact (when they believe the public/taxpayers aren't listening). I heard one of the leading black-hole researchers say just that last month at a talk on a remote observatory mountain.
4. *Personal fascination:* My biggest fascination is that people don't get tired of being presented the "final" proof of the existence of black holes every few months.

Kip Thorne, a professor of theoretical physics at the California Institute of Technology and author of *Black Holes and Time Warps*, wrote to me:

> *Unsolved problem:* The greatest problem about black holes today, I believe, is the precise nature of the singularity in their cores; to understand it we must first understand quantum gravity.

From Werner Benger, a student of astronomy and physics at the University of Innsbruck, Austria:

1. *Misconceptions:* That black holes do exist.
2. *Public fascination:* Black holes are an extreme, but relatively simple, case of an application of general relativity. They show some of the fundamental concepts in a way that a lot of people may understand. Black holes also show a physical-based path to infinity. Black holes are esoteric objects that may be studied by mathematical formalism. Such things are fascinating.

3. *Useful function:* A courageous expedition could plan to go near a collapsed star surface to benefit from the time-dilation. Such an object could therefore be used as a time-machine, but only for one direction: to the future.

Alexander Honcharik is a design engineer for Intel Corporation in Phoenix, Arizona. One of his hobbies is amateur astronomy. He says:

1. *Public fascination:* Black holes are generally thought of as big, powerful, and mysterious.
2. *Useful function:* I've thought that it might be a neat idea to use a black hole as a toxic waste disposal. You'd somehow stabilize the position of a small black hole (a few meters in circumference perhaps) and feed it with matter at the same rate that it evaporates due to Hawking radiation. However, I wouldn't want to be in the same solar system where this disposal system was being tested.
3. *Unsolved problem:* I'd like someone to find a method for detecting black holes (and other dark matter) other than by observing gravitational effects. This doesn't seem likely.
4. *Personal fascination:* I am fascinated by black holes for the same reason everyone else is. Again, they're big, powerful, and mysterious. In addition, they allow one to imagine many interesting scenarios: cosmic waste disposal units, time travel through wormholes, the outlandish and grotesque stretching of the human body as it approached a black hole of appropriate mass.

From Sylvan Jacques, a retired Ph.D. physicist who studied plasmas, magneto-hydrodynamics, the solar system, classical field theory, and relativity of continua:

1. *Public fascination:* Whoever chose the term *black holes* was a public relations genius. I think researchers first called these objects "massive singularities." What if they had been called MSOs or SIMOs (singular massive objects)? The term *black holes* is sexy. The famous scientific phrase, "Black holes have no hair," is also pretty suggestive. The idea of infinitely dense objects with gravity so strong that light can't escape is also very attractive. I think these

reasons are largely responsible for why black holes are more popular than quasars, pulsars, or neutron stars. The fact that black holes represent astronomical extreme or limit case makes them of particular interest.

I close this chapter with a relevant quote from John Wheeler, the scientist who coined the term *black hole*:

> The advent of the term *black hole* in 1967 was terminologically trivial but psychologically powerful. After the name was introduced, more and more astronomers and astrophysicists came to appreciate that black holes might not be a figment of the imagination but astronomical objects worth spending time and money to seek.

Author's Musings

These Notes *as I see them, relate not to lectures but to feeling. I'm sure my readers differ from me on many things, but I hope that we share the essence of wonder and longing for what we may never quite understand.*

— Piers Anthony, 1991, *Virtual Mode*

Piers Anthony, one of science fiction and fantasy's most prolific talents, established the interesting practice of placing an "Author's Note" section at the end of some of his books. In these notes, he gives that slice of his life occurring during the writing of a current novel, complete with discussions of social issues and unfinished thoughts. In keeping with Piers, I started this practice in my book *Chaos and Wonderland* and continue it here by providing you a slice of some of the mail I have received while writing *Black Holes: A Traveler's Guide*. Also included are miscellaneous recent thoughts and breakthroughs in our understanding of black holes and reader responses to problems posed in my previous books.

Theorists Attempt to Eliminate Black Holes

I don't believe in black holes.
—Philip Morrison, *Massachusetts Institute of Technology*

Many theorists continue to be uneasy with points of infinite density and gravity, or singularities, that lurk in the heart of black holes. The December 23, 1994, issue of *Science* reports that a handful of physi-

153

cists think black-hole seekers are pursuing a chimera, something like the ether believed to fill space in the nineteenth century. In fact, some researchers have attempted to reformulate Einstein's mathematics in attempt to remove the singularity and make black holes "disappear." These physicists believe the astronomical evidence for black holes is merely evidence for dense objects such as neutron stars or clusters of giant stars.

So far, the theorists' ideas have been greeted with skepticism, part of the continual parade of useful attempts to modify general relativity in such a way as to remove the singularity while keeping other well-tested aspects of Einstein's theories.

For more information: Flam, F. (1994). Theorists make a bid to eliminate black holes. *Science*, 266(5193): 1945. Also see: Horgan, J. (1995). Bashing black holes: Theorists twist relativity to eradicate an astronomical anomaly. *Scientific American*, July, 273(1): 16.

Unraveling the Universe

Cosmology has only lately crossed the dividing line from theology into true science.
—M. Lemonick and J. Nash, *Time*, March 6, 1995

The years 1994 and 1995 were banner years for astronomers asking questions such as: What is the age of the universe? What is the universe made of? What is the fate of the universe? And how is the cosmos structured? Due in part to images received from the Hubble space telescope, astronomers have put forth a bewildering array of conflicting theories. For example, for decades scientists believed the universe to be about 15 to 20 billion years old, but new images suggest the age is somewhere between 8 billion and 12 billion years, which seems to make the universe much younger than some of the stars it contains. As I write this, no one is sure how to best interpret the conflicting findings. Most astronomers today believe that the universe is between 8 billion and 25 billion years old, and has been expanding outward ever since. The universe seems to have a fractal nature with galaxies hanging together in clusters. These clusters form larger clusters (clusters of clusters). "Superclusters" are clusters of these clusters of clusters.

In recent years, there have been other baffling theories and discoveries. Here are just a few:

- In our universe exists a Great Wall consisting of a huge concentration of galaxies stretching across 500 million light-years of space.
- In our universe exists a Great Attractor, a mysterious mass pulling much of the local universe towards the constellations Hydra and Centaurus.
- There are Great Voids in our universe. These are regions of space where few galaxies can be found.
- Inflation theory continues to be an important theory describing the evolution of our universe. Inflation theory suggests that the universe expanded like a drunken balloon-blower's balloon while the universe was in its first second of life.
- The existence of dark matter also continues to be hypothesized. Dark matter consists of subatomic particles that may account for most of the universe's mass. We don't know of what dark matter is composed, but theories include neutrinos (subatomic particles), WIMPs (weakly interacting massive particles), MACHOs (massive compact halo objects), or black holes.
- Cosmic strings and cosmic textures are hypothetical entities that distort the space-time fabric.

For more information, see: Lemonick, M., and Nash, M. (1995). Unraveling universe. *Time*, March 6, 145(9): 77–84.

A Close Look at a Galaxy's Center

Astronomers have always been eager to look at the center of active galaxies called quasars that shine so brightly that a black hole is thought to power them by dragging in nearby matter. By monitoring how fluctuations in light echo from the center of the active galactic nucleus (AGN) in galaxy NGC 5548—off clouds of gas whirling around the nucleus at a distance of a few light-days, scientists have found support for the idea that a black hole's gravity is the AGN's driving force. Observations made using ground-based telescopes, and satellites including the Hubble space telescope, suggest that the center of NGC 5548 is powered by a 20 million solar mass black hole. For more information, see: Travis, J. (1995). A closer look at an active galaxy's engine, *Science*, March, 267(5205): 1768–1769.

Astronomers have long suspected that an unseen object with one million times the mass of the sun is pulling in stars and gas at the center of our Milky Way galaxy. However, the fact that this radio source emits too little radiation (gamma rays and X rays) has also cast some doubt on this theory. To quell doubts, recent models show that gas can enter a black hole before it has a chance to emit much radiation. Therefore, a black hole could gobble gas at our galaxy's center yet remain faint. For more information, see: Kaiser, J. (1995). Does the Milky Way hide its black hole? *Science News*, April, 147(15): 230.

Black Holes and String Theory

In the past few years, theoretical physicists have been using mathematical constructs called "strings" to explain all the forces of nature, from atomic to gravitational. String theory describes elementary particles as vibrational modes of infinitesimal strings that exist in ten dimensions. How could such things exist in our four-dimensional space-time universe? String theorists claim that six of the ten dimensions are "compactified": tightly curled up (in structures known as Calabi-Yau spaces) so the extra dimensions are essentially invisible. Unfortunately, there are so many different ways to create universes by compactifying the six dimensions that string theory is difficult to relate to the real universe. In 1995, researchers suggested that if string theory takes into account the quantum effects of charged mini black holes, the thousands of four-dimensional solutions may collapse to only one. Tiny black holes, with no more mass than an elementary particle, and strings may be two descriptions of the same object. Thanks to the theory of mini black holes, physicists now hope to mathematically follow the evolution of the universe and select one particular Calabi-Yau compactification—a first step to a testable "theory of everything." (For more information, see: Taubes, G. (1995). How black holes may get string theory out of a bind. *Science*, June, 268(5218): 1699.)

Arthur C. Clarke

I have received a tremendous number of letters regarding topics in my previous books. For example, in *Keys to Infinity* and *Chaos in Wonderland* I mention the extraordinary π computing abilities of Johann Dase. Arthur C. Clarke recently wrote to me that he simply doesn't believe the story of Dase calculating π to 200 places in his head. Clarke says, "Even though I've seen fairly well authenticated reports of other incredible feats of mental calculation, I think this is totally beyond credibility." Clarke, stimulated by my Dase report, recently wrote Stephen Jay Gould asking how it is possible for such extraordinary abilities as human calculators to have evolved through natural selection. Clarke says, "What is the survival value in the jungle of the ability to multiply a couple of 50-digit numbers together?"

Here is some background on Dase. In 1844, Johann Martin Zacharias Dase (1824–1861), a human computer, is said to have calculated π correct to 200 places in less than two months:

$\pi =$ 3.14159 26535 89793 23846 26433 83279 50288 41971 69399
37510 58209 74944 59230 78164 06286 20899 86280 34825 34211
70679 82148 08651 32823 06647 09384 46095 50582 23172 53594
08128 48111 74502 84102 70193 85211 05559 64462 29489 54930
38196.

To compute π Johann Dase supposedly used: $\pi/4 = \arctan(\frac{1}{2}) + \arctan(\frac{1}{5}) + \arctan(\frac{1}{8}) \ldots$ with a series expansion for each arctangent. Dase ran the arctangent job in his brain for almost two months.

I would be interested to hear from readers who can confirm or deny this story.

Here's another tidbit that Arthur C. Clarke sent me to ponder: "If there are infinitely many digits in π, then π must contain every possible series of numbers. Therefore, it contains π—all its digits, And *that* subset must contain π, so on for ever and ever. . . . This way lies madness."

Arthur C. Clarke was also particularly interested in the chapter "Ladders to Heaven" in my book *Keys to Infinity*. In this chapter I describe a ladder that stretches from the Earth to the moon and ask readers to speculate on how long it would take a human to climb the ladder. Clarke wrote to me, "Cliff, I'm surprised that in the chapter 'Ladders to Heaven' you didn't mention the extensive literature on the 'space elevator'—including my own *Fountains of Paradise*. We now have material to build it as, a couple of years ago, a group of chemists at Rice University, Houston, announced the discovery of the tubular form of C-60, and in their press release they specifically referred to it as a building material for the space elevator."

IQ-Block

In *Chaos in Wonderland* I mentioned a puzzle called IQ-Block (manufactured by Hercules) consisting of ten polygonal pieces that fit together to form a square. I asked in how many different ways could the pieces be assembled to form a square. (The manufacturer boasted that there were more than 60 solutions.)

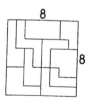

Charles Ashabacher, book editor of the *Journal of Recreational Mathematics* wrote to me that his computer program found more than 1,000 solutions rather quickly. He believes the number of solutions is in the tens of thousands. In my book I asked, "Can a square be constructed if one piece is removed?" It appears impossible to form a square after one piece is removed, but Charles also wrote to me that he found two ways to create a square after *two* pieces are removed. His constructions were found via a computer, and it seems likely there are others.

Latööcarfians

Dennis Gordon from Madison, Wisconsin, writes: "Dear Dr. Pickover, *Chaos in Wonderland* was a lot of fun to read, but how do you pronounce öö in your creatures called the Latööcarfians. The best a few of us could come up with was a hiccup sound as in Buddy Holly song."

The öö is produced in the same way you would produces a German ö umlaut. Ask a German friend.

So far no reader has guessed at the hidden meaning behind the word *Latööcarfians*.

Paul Hartal writes of *Chaos in Wonderland*, "On page 181 there seems to be an amazing case of synchronicity: the writing which is described as looking 'vaguely like Hebrew' is actually a description of π featured in the Bible."

Double Apocalyptic Powers

In my book *Chaos in Wonderland*, an extraterrestrial race of philosophers presents the following number to his disciples and asks them what is special about it:

182687704666362864775460604089535377456991567872

After much discussion, one young insectile student speaks: "It is the first power of two that exhibits three consecutive 6's."

The audience of philosophers applauds with delight. In fact the large number with "666" in its digits is equal to 2^{157}. I've called numbers of the form 2^i, which contain the digits 666, "apocalyptic powers" because of the prominent role 666 plays in the last book of the New Testament. In this book, called The Revelation (or Apocalypse) of John, "666" is designated by John as the Number of the Beast, the Antichrist—and various mystics have devoted much energy to deciphering the meaning of 666. (More recently, mystical individuals of the extreme fundamentalist right have noted that each word in the name Ronald Wilson Reagan has six letters. Interestingly, the book called the Revelation (or Apocalypse) of John is the last book of the New Testament, with the exception of the Syriac-speaking church, which has never accepted it.)

In *Chaos in Wonderland*, I also asked readers if there exist "double apocalyptic powers," which contain six 6's in a row. At the time I wrote the book, none were known.

On August 28, 1994, Werner Knoeppchen of Glenwood Springs, Colorado, sent me a page from a printout of the number $2^{5,000,000}$. Werner writes, "The number contains six 6's in a row. Therefore it is an apocalyptic power. I do not know if it is the lowest. The printout for $2^{5,000,000}$ is over 500 pages long, and the number contains 1,505,150 digits. It required two weeks for a Mac IICI to calculate this number running Mathematica."

Werner's double apocalyptic power contains

"... 10556666660670 ...,"

which he proudly circled in red ink.

Charles Ashbacher of Cedar Rapids, Iowa, wrote a Pascal program that searched for double apocalyptic powers. He found such powers with exponents of i as follows: 2269, 2271, 2868, 2870, 2954, 2956, 5485, 5651, 6323, 7244, 7389, 8909, 9195, 9203, 9271, 9273, 9275, and 9514. (Why are there several "twins" that differ by two: 2269 and 2271, 2868 and 2870, 2954 and 2956? Why should a "triplet" exist: 9271, 9273, and 9275?)

Christopher Becker from Homer, New York, used a DEC VAX 6410 and verified Ashbacher's findings regarding double apocalyptic powers. Becker notes that the first such number 2^{2269} has 684 digits and has "666666" at the 602nd position. For *single* apocalyptic powers, he finds $2^{157}, 2^{192}, 2^{218}, 2^{220}$, and 2^{222}. Curiously, 2^{666} is itself an apocalyptic power. Between 2^{2000} and 2^{3000} he finds that more than half of the exponents are apocalyptic powers. Becker has also searched for St. John powers that have the digits "153" (Simon Peter caught 153 fish for Jesus). 2^{115} is the first St. John power.

Christopher Becker later used a DEC Alpha computer to search for *triple* apocalyptic numbers with nine 6's in a row. He searched as high as 2 raised to a quarter million using his custom C program. After using five hours of computing time, he found the following triplet of triple apocalyptic exponents less than 250,000 that differ by two: 192916, 192918, and 192920. He also found 212253, 237373, 241883, and 242577.

John Graham of Penn State Wilkes-Barre, Pennsylvania, and R. W. W. Taylor of the National Technical Institute for the Deaf (Rochester Institute of Technology, Rochester, New York) have both proven that there is an infinite number of apocalyptic powers. Proofs are available upon request.

Friden Calculators

In *Mazes for the Mind* I mention Friden calculators, electromechanical machines popular in the 1950s. This brought a barrage of mail from readers reminiscing about these chunky but useful calculators. For example, Dennis Gordon from Madison, Wisconsin, writes:

> We had an older Friden calculator in our lab when I was a graduate student, and our favorite way of unnerving a coworker was to multiply 9,999,999,999 by 9,999,999,999 and then quickly leaving the room. The noise was terrific and lasted for several minutes. My father had such a machine on his desk at work when I was a kid, and I thought it was something from the far distant future. When I purchased my first electronic calculator, the first thing I did was to perform the same calculation, and I was very disappointed this multiplication was no different an experience from any other multiplication.

Toilet Paper

Several readers wrote to me about two unusual chapters in *Mazes for the Mind* titled "Toilet Paper and the Infinite" and "Squashed Archimedean Model of a Hyper-Toilet Roll of Paper." Dennis Gordon writes:

Considering the Archimedean model of a roll of toilet paper, on page 178, I think I arrived at a different value for the length. Looking at the drawing (Figure 34.1 in *Mazes for the Mind*), as I did while having a beer with some other patrons at my favorite tavern (the 602 Club, and maybe we even generated a few books sales for you), we took your suggestion and tried to guess the length of the line. Somehow 123,000 inches (about 2 miles) seemed high. I thought that the spacing between the lines (white space, or thickness of the paper) looked like a familiar distance, so when I got home I measured it with a printer's ruler and found it to be a very nice 1/36 inch. The radius was 163/72 inches. Using a value of $r_2 = 163/72$ inches and $r_1 \sim 0$, I find $l \sim 759$ inches or about 48 inches.

Chernikov Pattern Puzzle

In *Mazes for the Mind*, I conducted an experiment using a Chernikov design. My goal was to test the ability of words to describe a graphic diagram. I started by limiting myself to a 75-word description of a Chernikov pattern. Using only this description and no illustration, I asked people to draw the target design:

> There are four quarter-sized circles arranged so they touch. They are on top of one larger circle with a thick edge, and cover the circle slightly. Inside each of the four circles is a black dot which touches the edge of each of the circles. The four circles also contain six circles inside them. The six circles get smaller and smaller, but all of their edges touch the black dot.

Notice that I also limited myself to nonmathematical terms when describing the target picture.

I received many drawings from readers who attempted to draw the figure. A few respondents noted that my English-language description was ambiguous. Some people asked me, "What is the diameter of the large circle?" "What does 'cover slightly' mean?" "How big is the black dot?" "Are the groups of circles nested or adjacent?" I could only respond to these critical people that the description was the best I could do with 75 words and without resorting to mathematical jargon.

I received numerous alternative descriptions from readers including medical doctors, oceanographers, and lawyers. The following is one description from Katie Traverso, age 13. She writes, "I thought that it was a very interesting idea and decided to follow up on what you said about different people having different ways of describing the target figure." Here is her description. What figure would you have drawn based on her description?

> There is a thick-edged circle. Four circles are arranged so one is slightly overlapping the thick circle at the top, another at the bottom, the other two at the sides. They are large enough to touch sides. Inside each circle is a black dot that touches the edge facing the middle. Six circles are inside each circle. Their edges touch the side of the circle that the black dot does. Each circle gets smaller.

Visit the Bizarre Chess Leapers

Ed Pegg and colleagues were stimulated by the various chess variations described in *Mazes for the Mind*. What happens if new pieces are introduced that move in unusual ways? For example, in chess, the knight makes an L-shaped move of one square in one direction and two squares in a perpendicular direction over squares that may be occupied. As such, the knight may be called a "one-two leaper" or "(1, 2) leaper." What other kinds of leapers can be used in chess? Ed has written me the following, and I thank him for giving me permission to reproduce his letter here.

A *tour* of the chessboard by a knight occurs when the knight visits each square just once. The famous Swiss mathematician Leonhard Euler (1707–1783) studied these tours extensively. However, there are strange and exotic leapers to consider in addition to the traditional knight. For example, the *Zebra* is a (2, 3) leaper. The *Giraffe* is a (1, 4) leaper. Can either of these pieces, or any other (*m*, *n*) leaper, make a tour of any board of any size? Here are some solutions:

1	26	45	68	83	16	5	24	31	14
28	47	74	43	70	3	22	53	72	33
67	84	55	0	25	30	13	82	17	6
44	69	2	27	46	73	32	15	4	23
75	42	29	48	85	54	71	34	21	52
56	99	66	93	58	79	18	7	12	81
97	90	39	64	95	20	9	88	37	60
92	49	76	41	62	11	86	51	78	35
65	94	57	98	89	38	59	80	19	8
40	63	96	91	50	77	36	61	10	87

5	66	19	54	75	42	29	38	3
76	43	28	37	4	65	20	55	74
23	14	71	62	49	34	79	10	25
48	33	80	9	24	15	70	61	50
67	6	53	18	45	30	41	2	39
44	77	36	27	12	21	64	73	56
13	22	63	72	57	78	35	26	11
58	47	32	81	8	69	16	51	60
7	68	17	52	59	46	31	40	1

Closed Zebra tour of 10x10 found by John Scholes

Open Giraffe tour of 9x9 found by Juha Saukkola

8	37	58	47	30	5	36	61
45	52	21	2	13	44	51	28
64	15	24	41	54	33	16	25
39	56	49	18	9	38	57	48
12	31	4	27	62	11	22	3
7	34	59	46	29	6	35	60
42	53	20	1	14	43	50	19
63	10	23	40	55	32	17	26

49	12	51	16	41	18	45	30
40	19	44	29	48	9	62	15
59	8	63	2	39	20	35	26
38	23	32	25	58	5	56	3
11	50	13	52	17	42	31	46
*	43	28	47	10	61	14	53
7	60	1	54	21	36	27	34
22	37	24	33	6	57	4	55

Closed Pythagorean tour of 8x8 found by Juha Saukkola

Almost Giraffe tour of 8x8 found by Juha Saukkola

("Pythagorean" = (3,4) Leaper (Antelope) + (0,5) leaper)

Dan Cass has proven that an 11-by-11 Zebra tour (open or closed) is impossible. Proving the Zebra cannot tour the 8-by-8 or 9-

by-9 boards is an interesting avenue of research. Ed Pegg, Jr., has proven that a 12-by-12 Zebra tour is impossible. Juha Saukkola is almost certain that the 8-by-8 Giraffe tour is impossible. Donald Knuth has looked at the minimum boards needed for a (m, n) leaper to travel from one cell to any other cell. Such traveling is possible only if m and n are relatively prime and of different parity. The smallest board is $m+n$ by $2 \times (min(m, n))$.

Here are some unsolved problems:

1. Do tours exist for the following leapers? $(1, 6)$, $(2, 5)$, $(2, 7)$, $(3, 4)$, $(3, 6)$, $(4, 5)$, $(4, 7)$, $(5, 6)$, $(6, 7)$
2. Is there an open Zebra tour for the 9-by-10 board? (The closed tour was proven impossible by Ed Pegg, Jr.)
3. Is there a Giraffe tour on a board smaller than 9-by-9?
4. For a general (m, n) leaper, what is the smallest board it can tour?
5. Is there an (m, n) leaper (such that $GCD(m, n) = 1$ and $m+n = odd$) that cannot tour a board of any size?

Updates

Frequently I receive letters with suggestions regarding my previous books. For example, Mark Plus informs me that the address for the Extropy organization is now out of date. The current address is: Extropy Institute, 13428 Maxella Ave, #273, Marina Del Rey, CA 90292. *Extropy* is an unusual journal devoted to nanotechnology, life extension, artificial life, digital economies, and related.

Michael Strasmich of the Great Media Company (formerly Media Magic) notifies me that his new address is PO Box 598, Nicasio, California 94946. His fine company distributes books, videos, prints, and calendars relating to computers in science and art. Catalog available.

Here are some other fascinating organizations, newsletters, and magazines.

1. *bOING-bOING Magazine.* Unusual newsletter covering topics such as cyberspace, data encryption, chaos, artificial life, cryonics, nanotechnology, self-organization, game theory, evolution, brain toys, programming, alien life, psychedelia. Very weird short sto-

ries. Fringe culture. Cyberpunk. A "perpetual novelty brain jack designed to demolish the gray shield of consensus reality." Sample, $3.95; four issues, $14. 4500 Forman Ave #2, Toluca Lake, CA 91602.

2. *QUANTUM.* Beautiful, well illustrated, glossy, student magazine of math and science. Highly recommended. Conveys the breadth and wonder of math and physics. Contact: Springer-Verlag New York, Inc., Attn: Journal Promotion Department, 175 Fifth Ave., New York, NY 10010.

3. *AMYGDALA*, a fascinating newsletter on fractals. Write to AMYGDALA, Box 219, San Cristobal, NM 87564, for more information.

4. *Powell's Technical Bookstore Newsletter.* 33 NW Park Ave., Portland, OR 97209. (Computing, electronics, engineering, and science books.)

5. *YLEM—Artists using science and technology.* This newsletter is published by an organization of artists who work with video, ionized gases, computers, lasers, holograms, robotics, and other nontraditional media. It also includes artists who use traditional media but who are inspired by images of electromagnetic phenomena, biological self-replication, and fractals. Contact: YLEM, Box 749, Orinda, CA 94563.

6. *Recreational and Educational Computing Newsletter.* Dr. Michael Ecker, 909 Violet Terrace, Clarks Summit, PA 18411. (Devoted to the playful interaction of computers and mathemagic—from digital delights to strange attractors. Puzzles, program teasers.)

7. *ART MATRIX*, creator of beautiful postcards and videotapes of exciting mathematical shapes. Write to ART MATRIX, P.O. Box 880, Ithaca, NY 14851, for more information.

8. *Small Computers in the Arts Newsletter.* An organization that supports artists working with small computers. 5132 Hazel Ave., Philadelphia, PA 19143.

9. *Fractal Report*, a great newsletter on fractals. Published by J. de Rivaz, Reeves Telecommunications Lab. West Towan House, Porthtowan, Cornwall TR4 8AX, United Kingdom.

10. *Strange Attractions*. A store devoted to chaos and fractals (fractal art work, cards, shirts, puzzles, and books). For more information, contact: *Strange Attractions*, 204 Kensington Park Road, London W11 1NR England.

11. *Astronomy Society Catalog*. Explore the cosmos with slides, videos, books, software, and gifts. The nonprofit *Astronomical Society of the Pacific* has served as a bridge between astronomers and the public since 1889. With members in 50 states and more than 75 countries, the society is an important resource for astronomy teachers, students, and enthusiasts. Their catalog sells magnificent color posters of Jupiter and Saturn (with their satellites), various nebulae, galaxies, other planets, and moons. Also listed are giant moon maps, comet posters, spectra, and astronomical image processing software with 19 sample images of planets and stars (MS-DOS). Also: interactive planetarium software for the Mac, Amiga, and IBM compatibles, and "Greetings from Outer Space" tourist cards from the cosmos. Members of the society receive a magazine, monthly sky calendars, star maps, etc. Contact: Astronomical Society of the Pacific, 390 Ashton Avenue, San Francisco, CA 94112.

12. *International Association for Astronomical Arts*, for anyone interested in astronomical "space" art. The IAAA is the first and only guild of astronomical artists. Originally conceived by a small core of professionals, they have grown to include student members and associate members including agents, collectors, planetariums, and writers. Their goal is to nurture the field of space art, providing a forum for the exchange of ideas and information. Contact: IAAA, 4160 Willows Road, Alpine CA 91901. In Europe, contact: IAAA, 99 Southam Road, Hall Green, Birmingham B28 0AB England.

13. *HyperSpace*, a fascinating journal on all subjects relating to higher dimensional geometries, geometry and art, and unusual patterns. The journal has articles in English and Japanese. Contact: Japan Institute of Hyperspace Science, c/o K. Miyazaki, Graduate School of Human and Environmental Studies, Kyoto University, Sakyo-ky, Kyoto 606 Japan.

14. *Odyssey*, a colorful astronomy magazine especially for students. Contact: Odyssey, Cobblestone Publishing, 7 School Street, Peterborough, NH 03458.

Pieces of Pi

On page 248 of *Chaos in Wonderland* I asked whether the expression π^2 ever occurs in geometry or physics. Mike Lawrence of Tucson, Arizona, says such a term is found in the equation that describes the throughput T of a plano-convex lens/detector system: $T = G\pi^2 (D_1/2)^2$. Here, $G = 0.5(Z - \sqrt{Z^2 - 4X^2Y^2})$, $X = D_2/(2f)$, $Y = 2f/D_1$, and $Z = 1+(1+X^2)Y^2$. D_1 is the lens diameter, f is the lens focal length, and D_2 is the detector length.

Smorgasbord for Computer Junkies

How to Calculate the Mass of a Black Hole (Chapter 1)

C Program Code

```
/* Compute Mass of Black Hole */
/* Print Data for M vs C vs P Plot        */
#include <math.h>
#include <stdio.h>

main()
{
     float CO, /* Orbital circumference, km */
           PO, /* Orbital Period, seconds    */
           G; /* Gravitational constant km**3/s**2 per solar
mass*/
     float pi, MassHole;

     G = 1.327E11;
     pi = 3.14159;

     for(CO = 1E5; CO < 1E7; CO=CO+2E5){
          printf("CO is: %f\n",CO);
          for(PO = 10; PO < 1000; PO=PO+10){
              MassHole = CO*CO*CO/(2*pi*G*PO*PO);
              printf("%f %f\n",PO, MassHole);
          }
     }
}
```

BASIC Program Code

```
10 REM Compute Mass of Black Hole
15 REM Print Data for M vs C vs P Plot
20 REM CO is the orbital circumference, km
```

```
30 REM PO is the orbital period, seconds
40 REM G is the Gravitational constant, km**3/s**2 per solar mass
50 G = 1.327E11
60 P = 3.14159
70 FOR CO = 1E5 TO 1E7 STEP 2E5
80    PRINT "CO is:"; CO
90    FOR PO = 10 TO 1000 STEP 10
100       M = CO*CO*CO/(2*P*G*PO*PO)
110       PRINT PO, M
120    NEXT PO
130 NEXT CO
140 END
```

How to Compute the Circumference of the Event Horizon (Chapter 2)

C Program Code

```
/* Compute Circumference of Event Horizon */
#include <math.h>
#include <stdio.h>

main()
{
    float c, /* The speed of light, km/s */
          m, /* Black hole's mass, solar masses */
      circum, /* Black hole's circumference, km */
          G; /* Gravitational constant, km**3/s**2 per solar mass */
    float pi;

    G = 1.327E11;
    pi = 3.14159;
    c = 2.998E5;
    for(m = 1; m <= 500; m=m+100) {
        circum = 4*pi*G*m/(c*c);
        printf("Hole Mass: %f solar masses\n",m);
        printf("Circumference: %f kilometers\n",circum);
      }
}
```

BASIC Program Code

```
10 REM Compute Circumference of Event Horizon
20 REM S is the speed of light, km/s
40 REM G is the Gravitational constant, km**3/s**2 per solar mass
45 REM M is Black hole's mass, solar masses
50 G = 1.327E11
60 P = 3.14159
```

```
70 S = 2.998E5
80 FOR M = 1 TO 500 STEP 100
90    C = 4*P*G*M/(S*S)
100   PRINT "Hole Mass:"; M; "solar masses";
105   PRINT "Circumference:"; C; "kilometers"
110 NEXT M
140 END
```

How to Calculate Black Hole Tidal Forces (Chapter 3)

C Program Code

```c
/* Compute Tidal Force Upon Body */
#include <math.h>
#include <stdio.h>

main()
{
    float l, /* The distance between two points, km */
          m, /* Black hole's mass, solar masses    */
      circum, /* Black hole's circumference, km */
          a, /* difference in acceleration, km/sec**2 */
          G; /* Gravitational constant, km**3/s**2 per solar mass */
    float pi;

    G = 1.327E11;
    pi = 3.14159;
    circum = 100000;
    m = 303;
    for (l=0.0018; l<=.004; l=l+0.001) {
        a = 16*pi*pi*G*m*l/(circum*circum*circum);
        /* Convert km/s**2 to g units   */
        a = (a*1000)/9.81;
      printf("Length: %f km, Tidal Force: %f g\n",
             l,a);
    }
}
```

BASIC Program Code

```basic
10 REM Compute Tidal Force Upon Body
20 REM G is the Gravitational constant, km**3/s**2 per solar mass
30 REM M is the black hole's mass, solar masses
40 REM L is the distance between two points, km
50 REM C is the circumference, km
60 REM A is difference in acceleration, km/sec**2
70 C = 100000
80 G = 1.327E11
```

```
90 P = 3.14159
100 M = 303
110 FOR L = .0018 TO .004 STEP .001
120     A = 16*P*P*P*G*M*L/(C*C*C)
130     REM Convert km/s**2 to g units
140     A = (A*1000)/9.81
150     PRINT "Length: "; L; "km, Tidal Force: "; A; "g"
160 NEXT L
170 END
```

How to Calculate an Accelergram (Chapter 3)

C Program Code

```c
/* Compute Accelergram */
#include <math.h>
#include <stdio.h>

main()
{
  float dist, theta, circum, r, x1, x2, y1, y2, pi;
  int i;
  pi = 3.14159;
  srand(1234567);

  for (i=0; i<10000;i++) {
    /* Generate random circumference */
    circum = 2.*pi*(float)rand()/32767.;
    theta  = 2.*pi*(float)rand()/32767.;
    r = circum/(2.*pi);
    x1 = r*cos(theta);
    y1 = r*sin(theta);
    /* Determine length of vector */
    r = r - .001/(circum*circum*circum);
    if (r<0) r=0;
    x2 = r*cos(theta);
    y2 = r*sin(theta);
    dist = sqrt((x2-x1)*(x2-x1) + (y2-y1)*(y2-y1));
    /* Don't plot tiny vectors too small to see */
    if (dist > .001) {
        /* print end-points of each vector */
        printf("%f %f\n",x1,y1);
        printf("%f %f\n",x2,y2);
    }
  }
}
```

BASIC Code

```
10 REM Compute Accelergram
20 P = 3.14159
30 FOR I = 0 TO 1000
40    REM Generate random circumference
50    C = 2.*P*RND
60    REM Generate random angle
70    T = 2.*P*RND
80    R = C/(2.*P)
90    X1 = R*COS(T)
100   Y1 = R*SIN(T)
110   REM Determine length of vector
120   R = R - .001/(C*C*C)
130   IF R<0 THEN R=0
140   X2 = R*COS(T)
150   Y2 = R*SIN(T)
160   D = SQR((X2-X1)*(X2-X1) + (Y2-Y1)*(Y2-Y1))
170   REM Don't plot tiny vectors too small to see
180   REM Print end-points of each vector
190   IF D > .001 THEN PRINT X1;Y1
200   IF D > .001 THEN  PRINT X2;Y2
300 NEXT I
4000 END
```

Gravitational Lens Effect (Chapter 4)

BASIC Program Code

```
10 REM Compute Gravitational Lens Effect
20 REM C is circumference, units of black hole circum.
30 REM R is radius of disk containing light from universe
40 REM A is "angular diameter" of disk
50  P = 3.1415
60  PRINT "Circumference, angle, radius"
70  FOR I = 0 TO 2
80     IF I=2 THEN C = 1.10
90     IF I=1 THEN C = 1.05
100    IF I=0 THEN C = 1.01
110    A=300*SQR(1.0-1.0/C)
120    A=A/2
130    REM Assume holding diagram 12 inches from eye
140    R =12*TAN(A*P/180.)
150    PRINT C;A*2;R
160 NEXT I
170 END
```

C Program Code

```c
/* Compute shape of crushed universal disk, and draw circles
   filled with stars */
#include <math.h>
#include <stdio.h>

main()
{
  float circum,    /* circumference at which you hover above hole */
                   /* units of even horizon circumference */
        radius,    /* length in inches of universe portal */
        angle, pi, t, x, r, y, shift;
  int    i, j;
  pi = 3.14159; srand(1234567); shift=0;
  printf("Data for plotting circles:\n");
  for (i=0;i<3;i++) {
     if (i==2)  circum = 1.1;
     if (i==1)  circum = 1.05;
     if (i==0)  circum = 1.01;
     angle = 300*sqrt(1.0 - 1.0/circum);
     angle = angle/2.0;
     radius =12.0*tan(angle*pi/180.);
     printf("Circum, angle, radius: %f %f %f \n",
            circum,angle*2,radius);
     /* draw circle around crushed universe */
     for(t=0; t<=(2.*pi); t=t+.1) {
         x=radius*cos(t);
         y=radius*sin(t);
         /* Print points for a plot program */
         printf("%f %f\n",x+shift,y+shift);
        }
     /* Displace three plots so don't overlap */
     shift = shift + 2.5*radius;
    }
  /* Add random stars to circular field */
  shift=0;
  printf("Data for plotting stars:\n");
  for (i=0; i<3;i++) {
     if (i==2)  circum = 1.1;
     if (i==1)  circum = 1.05;
     if (i==0)  circum = 1.01;
     angle = 300*sqrt(1.0 - 1.0/circum);
     angle = angle/2.0;
     radius =12.0*tan(angle*pi/180.);
     for (j=0; j<=200;j++) {
         r=radius*(float)rand()/32767.;
         t=2.0*pi*(float)rand()/32767.;
         x=r*cos(t);
```

```
        y=r*sin(t);
        printf("%f %f\n",x+shift,y+shift);
    }
    /* Displace three plots so don't overlap */
    shift = shift + 2.5*radius;
  }
}
```

Compute Gravitational Blueshift (Chapter 5)

BASIC Program

```
5   REM Compute gravitational blueshift
10  REM C is ratio of circumference at which you hover above hole
15  REM   to the circumference of the event horizon
20  REM L1 is the wavelength of emitted light from star, meters
30  REM L2 is the wavelength of received light, meters
40  REM Yellow light is 5.8E-7 meters
50  L1 = 5.8E-7
60  I=0
70  PRINT "C/Ch  Lambda (emit)  Lambda(rec)"
80  FOR J=0 TO 20
90     C = 1.+0.5**I
100    I=I+1
110    L2 = L1*SQR(1.0 - 1/C)
120    PRINT C;L1;L2
130 NEXT J
140 END
```

C Program

```
/* Compute gravitational blueshift */
#include <math.h>
#include <stdio.h>
main()
{
    float circum,    /* circumference ratio */
          lambdas,   /* wavelength of emitted light from star,
                        meters */
          lambdar;   /* wavelength of received light, meters */
    int   i, j;

    lambdas = 5.8e-7; /* yellow light */
    i = 0;
    printf("C/Ch  Lambda (emit)  Lambda(rec)\n");
    for(j=0; j<=20; j++) {
        circum = 1+pow(0.5,i);
```

```
      i++;
      lambdar  = lambdas*sqrt(1.0 - 1./circum);
      printf("%f %g %g \n",circum,lambdas,lambdar);
   }
}
```

Gravitational Time Dilation (Chapter 6)

BASIC Program Code

```
10 REM Compute Gravitational Time Dilation
20 REM R is the ratio circum/circum(hole)
30 REM T2 is elapsed time, near hole
40 REM T1 is elapsed time, far away from hole
50 T1 = 1
60 I=0
70 PRINT "C/Ch     Time 1 (days)        Time2 (days)"
80 FOR J = 0 TO 20
90     R = 1+0.5**I
100    I=I+1
110    T2  = T1/SQR(1.0 - 1./R)
120    PRINT R, T1, T2
130 NEXT J
140 END
```

C Program Code

```
/* Compute Gravitational Time Dilation */
#include <math.h>
#include <stdio.h>
main()
{
   float ratio, /* circum/circum(hole) */
   time2, /* elapsed time, near hole */
   time1; /* elapsed time, far away from hole */
   int i, j;
   time1 = 1;
   i=0;
   printf("C/Ch  Time 1 (days)     Time2 (days)\n");
   for(j=0;j<=20;j++) {
      ratio = 1+pow(0.5,i);
      i++;
      time2  = time1/sqrt(1.0 - 1./ratio);
      printf("%f %f %f \n",ratio,time1,time2);
   }
}
```

Schematic Illustration of Rotating Black Hole (Chapter 7)

BASIC Program Code

```
10 REM Draw Schematic Illustration of Rotating Black Hole
20 REM Cross-Section
30 REM Compute ellipsoidal static limit shell
40 FOR T=0 TO 6.28 STEP 0.05
50    X=COS(T)
60    Y=0.7*SIN(T)
70    REM Print points for plot
80    PRINT X,Y
90 NEXT T
100 REM Compute Outer Event Horizon
110 FOR T=0 TO 6.28 STEP 0.05
120    X=0.7*COS(T)
130    Y=0.7*SIN(T)
140    REM Print points for plot
150    PRINT X,Y
160 NEXT T
170 REM Compute Inner Horizon Event
180 FOR T=0 TO 6.28 STEP 0.05
190    X=0.3*COS(T)
200    Y=0.3*SIN(T)
210    REM Print points for plot
220    PRINT X,Y
230 NEXT T
240 REM Compute Ring Singularity
250 FOR T=0 TO 6.28 STEP 0.05
260    X=0.1*COS(T)
270    Y=0.05*SIN(T)
280    REM Print points for plot
290    PRINT X,Y
300   NEXT T
310 REM Draw Axis of Rotation
320 PRINT "0 1"
330 PRINT "0 -1"
340 END
```

C Program Code

```c
/* Draw Schematic Illustration of Rotating Black Hole */
/* Cross-Section */
#include <math.h>
#include <stdio.h>
float theta,x,y;
main()
{
```

```
/* Compute ellipsoidal static limit shell */
for(theta=0; theta<6.28; theta=theta+.05){
    x=cos(theta);
    y=0.7*sin(theta);
    /* print points for plot */
    printf("%f %f\n",x,y);
}
/* Compute Outer Event Horizon */
for(theta=0; theta<6.28; theta=theta+.05){
    x=0.7*cos(theta);
    y=0.7*sin(theta);
    /* print points for plot*/
    printf("%f %f\n",x,y);
}
/* Compute Inner Horizon Event */
for(theta=0; theta<6.28; theta=theta+.05){
    x=0.3*cos(theta);
    y=0.3*sin(theta);
    /* print points for plot*/
    printf("%f %f\n",x,y);
}
/* Compute Ring Singularity */
for(theta=0; theta<6.28; theta=theta+.05){
    x=0.1*cos(theta);
    y=0.05*sin(theta);
    /* print points for plot*/
    printf("%f %f\n",x,y);
}
 /* Draw Axis of Rotation */
 printf("\nTABD\n");
 printf("0 1\n");
 printf("0 -1\n");
}
```

Embedding Diagrams for Stars (Chapter 8)

BASIC Program Code

```
10 REM Cross-Section of Embedding Diagrams for
20 REM Stars of Different Masses
30 REM R1 is star radius.  R2 is radius vector for plot.
40 REM M is mass.
50 R1 = 2
60 FOR M = 0.2 TO 1.0 STEP 0.1
70      FOR R2= -10 TO 10 STEP 0.2
80          REM CURVE INSIDE STAR
90          IF ABS(R2) <= R1 THEN Z = SQR((R1*R1*R1)/(2*M))
100         IF ABS(R2) <= R1 THEN Z = Z*(1 - SQR(1-2.*M*R2*R2/
            (R1*R1*R1)))
```

```
105          REM CURVE OUTSIDE OF STAR
110          IF ABS(R2) >= R1 THEN Z = SQR((R1*R1*R1)/(2*M))
120          IF ABS(R2) >= R1 THEN Z = Z*(1-SQR(1-2*M/R1))
130          IF ABS(R2) >= R1 THEN Z = Z+SQR(8*M*(ABS(R2)-2.*M))
140          IF ABS(R2) >= R1 THEN Z = Z-SQR(8*M*(R1-2*M))
156          REM DATA FOR PLOT
150          PRINT R2,Z
160      NEXT R2
170 NEXT M
180 END
```

C Program Code

```c
/* Cross-Section of Embedding Diagrams for
   Stars of Different Masses */
#include <math.h>
#include <stdio.h>
main()
{
        float M,   /* Mass of star */
              R,   /* Radius of star */
              r,   /* radial vector for plot */
              z;   /* amplitude of plot */
        R = 2;
        for (M=.2; M<=1.0; M=M+.1){
            for (r= -10; r <=10; r=r+.2){
                /* inside star */
                if (fabs(r) <= R)
                z = sqrt((R*R*R)/(2.*M)) * (1 -sqrt(1-2.*M*r*r/
                    (R*R*R)));
                /* outside star */
                if (fabs(r) >= R)
                z = sqrt((R*R*R)/(2.*M)) * (1-sqrt(1-2.*M/R))
                    + sqrt(8.*M*(fabs(r)-2.*M)) - sqrt(8*M*(R-2*M)) ;
                /* data for plot */
                printf("%f %f\n",r,z);
            }
        }
}
```

Embedding Diagrams for Black Holes (Chapter 8)

BASIC Program Code

```
10 REM Cross-Section of Embedding Diagram for Black Hole
20 REM R2 is radial vector for plot
25 M = 1
30 FOR R2= 2 TO 10 STEP 0.2
```

```
40              Z = SQR(8*M*(R2-2*M))
50              PRINT R2,Z
60 NEXT R2
70 FOR R2= 2 TO 10 STEP 0.2
80              Z = -SQR(8*M*(R2-2*M))
90              PRINT R2,Z
100 NEXT R2
110 REM LEFT PART OF BLACKHOLE
120 FOR R2= -10 TO 2 STEP 0.2
130             Z = SQR(8*M*(ABS(R2)-2*M))
140             PRINT R2,Z
150 NEXT R2
160 FOR R2= -10 TO 2 STEP 0.2
170             Z = -SQR(8*M*(ABS(R2)-2*M))
180             PRINT R2,Z
190 NEXT R2
200 END
```

C Program Code

```c
/* Cross-Section of Embedding Diagram for Black Hole */
#include <math.h>
#include <stdio.h>
main()
{

     float r,  /* radial vector for plot */
           z;  /* amplitude of plot */
     float M;
     M = 1;
     for (r= 2; r <=10; r=r+.2){
         z = sqrt(8*M*(r-2*M));
         /* print points for plotting */
         printf("%f %f\n",r,z);
     }
     for (r= -10; r <=-2;r=r+.2){
          z = sqrt(8*M*(fabs(r)-2*M));
          printf("%f %f\n",r,z);
     }
       for (r= -10; r <=-2;r=r+.2){
           z = -sqrt(8*M*(fabs(r)-2*M));
           printf("%f %f\n",r,z);
     }
       for (r= 2; r <=10;r=r+.2){
           z = -sqrt(8*M*(r-2*M));
           printf("%f %f\n",r,z);
     }
}
```

Compute Orbital Time Period of Binary Black Holes (Chapter 9)

BASIC Program Code

```
10 REM Compute Orbital Time Period of Binary Black Holes
20 REM D is the distance between two black holes, km
30 REM M is the black hole's mass, solar masses
40 REM P is the orbital period, seconds
50 REM G is the gravitational constant, km**3/s**2 per solar mass
60 G  = 1.327E11
70 P1 = 3.14159
80 M = 20
90 FOR D=5000 TO 100000 STEP 10000
100    P = 2*P1*SQR(D*D*D/(2.0*G*M))
120    PRINT D " kilometers", P " seconds"
130 NEXT D
140 END
```

C Program Code

```
/* Compute Orbital Time Period of Binary Black Holes */
#include <math.h>
#include <stdio.h>
main() {
    float dist, /* The distance between two black holes, km */
          m, /* Black hole's mass, solar masses     */
       period, /* Orbital period */
          G; /* Gravitational constant, km**3/s**2 per solar
               mass */
    float pi;
    G  = 1.327E11;
    pi = 3.14159;
    m = 20;
    for(dist=5000; dist<= 100000; dist=dist+10000) {
        period = 2*pi*sqrt(dist*dist*dist/(2.0*G*m));
        printf("%f kilometers, %f seconds\n",dist, period);
    }
}
```

Compute Time for Binary Holes to Coalesce (Chapter 9)

BASIC Program Code

```
10 REM Compute Time for Binary Holes to Coalesce
20 REM D is the distance between two black holes, km
30 REM M is the black hole's mass, solar masses
40 REM T is the time to coalesce
```

```
50 REM C is the speed of light, km/s
60 REM G is the Gravitational constant, km**3/s**2 per solar mass
70 G  = 1.327E11
80 P = 3.14159
90 C = 2.998E5
100 M=20
120 FOR D=5000 TO 100000 STEP 10000
130     T = (5.0/512.0)*((C*C*C*C)/(G*G*G))
140     T = T*((D*D*D*D)/(M*M*M))
150     REM convert from seconds to days
160     T = T/(60*60*24)
170     PRINT D " kilometers," T " days"
180 NEXT D
190 END
```

C Program Code

```
/* Compute Time for Binary Holes to Coalesce */
#include <math.h>
#include <stdio.h>
main()
{
    float dist, /* The distance between two black holes, km */
          m, /* Black hole's mass, solar masses     */
          time, /* time to coalesce */
          c, /* The speed of light, km/s */
          G; /* Gravitational constant, km**3/s**2 per solar
                mass */
    float pi,a,b;
    G  = 1.327E11;
    pi = 3.14159;
    c = 2.998E5;
    m = 20;
    for(dist=5000; dist<= 100000; dist=dist+10000) {
        time = (5.0/512.0)*((c*c*c*c)/(G*G*G))*
               ((dist*dist*dist*dist)/(m*m*m));
        /* convert from seconds to days */
        time=time/(60*60*24);
        printf("%f kilometers, %f days\n",dist, time);
    }
}
```

Radiation Received from Collapsing Star (Chapter 10)

C Program Code

```
10 REM Radiation received from collapsing star by distant observer
20 REM L is the luminosity of collapsing star
```

```
30 REM T is time
40 REM M mass of black hole
50 FOR M=20 TO 400 STEP 10
55     PRINT "Mass: ",M
60     FOR T=0 TO 500 STEP 10
70         L = EXP(-T/(3.0*SQR(3.0)*M))
75         REM Data for plot:
80         PRINT "Time: ",T,"Luminosity: ",L
90     NEXT T
100 NEXT M
110 END
```

BASIC Program Code

```c
/* Radiation received from collapsing star by distant observer */
#include <math.h>
#include <stdio.h>
main()
{
    float lumin,  /* luminosity of collapsing star */
          time,
            m;  /* mass of collapsing star */
    for (m=20; m<=400; m=m+10) {
        printf("Mass: %f\n",m);
        for(time=0; time<= 500; time=time+10) {
            lumin = exp(-time/(3.0*sqrt(3.0)*m));
            /* Data for plot: */
            printf("Time: %f Lumin: %f\n",time, lumin);
        }
    }

}
```

Compute Time to Die, Once Event Horizon Crossed (Chapter 11)

BASIC Program Code

```
10 REM Compute time to die, once event horizon crossed
20 REM M is mass of black hole, solar masses
30 REM T is time, seconds
40 FOR M = 20 TO 500 STEP 50
50     T = (1.54E-5)*M
60     PRINT "Mass: ";M;" solar masses,  Time: "; T; " seconds"
70 NEXT M
90 END
```

C Program Code

```c
/* Compute time to die, once event horizon crossed */
#include <math.h>
#include <stdio.h>
main () {
    float m, /* mass of black hole, solar masses */
       time; /* in seconds */
    for (m=20; m<=500; m=m+50) {
        time = 1.54e-5*M;
        printf("Mass: %f solar masses, Time: %f seconds\n",m,time);
    }
}
```

Quantum-Foam Embedding Diagram, 2-D (Chapter 12)

C Program Code

```c
/* Create quantum-foam embedding diagram, 2-D */
#include <stdio.h>
#include <math.h>
main()
{
    float r; /* random number, 0-1 */
    short c[513][513]; /* Arrays for holding 1 and 0 values */
    short ch[513][513];
    int size, time_steps, steps, i,j,k,sum;
    size = 80;   /* Use larger sizes for nicer images */
    /* Controls how many steps foam is to evolve */
    time_steps=20;
    /* Initially seed space with random 0's and 1's */
    for(i=0; i<=size; i++)
      for(j=0; j<=size; j++) {
         r = (float) rand()/32767.;
         if (r >= .5) c[i][j]=0; if (r <= .5) c[i][j]=1;
      }
    /* perform simulation based on twisted-majority rules
       to form foamlike objects in 2-D */
    for(steps=1; steps < time_steps; steps++) {
      for(i=1; i<size; i++){
         for(j=1; j<size; j++){
            /* compute sum of neighbor cells */
            sum = c[i+1][j+1] +
                  c[i-1][j-1] +
                  c[i-1][j+1] +
                  c[i+1][j-1] +
                  c[i+1][j] +
                  c[i-1][j] +
```

```
                    c[i][j+1] +
                    c[i][j-1] +
                    c[i][j];
                  if (sum == 9) ch[i][j]=1;
                  if (sum == 8) ch[i][j]=1;
                  if (sum == 7) ch[i][j]=1;
                  if (sum == 6) ch[i][j]=1;
          /* Notice "twist" in rules which destabilizes
             blob boundaries. */
          if (sum == 5) ch[i][j]=0;
          if (sum == 4) ch[i][j]=1;
          if (sum == 3) ch[i][j]=0;
          if (sum == 2) ch[i][j]=0;
          if (sum == 1) ch[i][j]=0;
          if (sum == 0) ch[i][j]=0;
        }
      }
    for(i=0; i<=size; i++) for(j=0;j<=size;j++)
      c[i][j]=ch[i][j];
  }
  /* If you like, you can make a movie of all frames */
  /* to show foam evolving.                          */
  printf("A plot of last frame of simulation\n");
  for(i=0; i<=size; i++){
    for(j=0; j<=size; j++){
      /* Crude attempt to draw blobs using characters */
      /* Better to use hi-res graphics package        */
      if (c[i][j] == 1) printf("*") ;else
        printf(" ");
    }
    printf("\n");
  }
}
```

BASIC Program Code

```
10 REM Create quantum-foam embedding diagram, 2-D
20 REM R is a random number, 0-1
30 REM C and H - Arrays for holding 1 and 0 values
35 DIM C(81,81)
40 DIM H(81,81)
50 REM Use larger size, S, for nice images
60 S = 80
70 REM T Controls how many steps foam is to evolve
80 T = 20
90 REM Initially seed space with random 0's and 1's
100 FOR I=1 TO S
120     FOR J=1 TO S
130         R=RND
```

```
140          IF R >= .5 THEN C(I,J) = 0
150          IF R <= .5 THEN C(I,J) = 1
160    NEXT J
170 NEXT I
175 REM Perform simulation based on twisted-majority rules
180 REM to form foamlike objects in 2-D
190 FOR K=1 TO T
200    FOR I=2 TO S-1
210       FOR J=2 TO S-1
220          REM compute sum of neighbor cells
230          A = C(I+1,J+1) + C(I-1,J-1) + C(I-1,J+1)
240          A = A+C(I+1,J-1) + C(I+1,J) + C(I-1,J) + C(I,J+1)
250          A = A+C(I,J-1) + C(I,J)
260          IF A = 9 THEN H(I,J) = 1
270          IF A = 8 THEN H(I,J) = 1
280          IF A = 7 THEN H(I,J) = 1
290          IF A = 6 THEN H(I,J) = 1
295          REM Notice "twist" in rules which destabilizes
267          REM blob boundaries.
300          IF A = 5 THEN H(I,J) = 0
310          IF A = 4 THEN H(I,J) = 1
320          IF A = 3 THEN H(I,J) = 0
330          IF A = 2 THEN H(I,J) = 0
340          IF A = 1 THEN H(I,J) = 0
350          IF A = 0 THEN H(I,J) = 0
360       NEXT J
370    NEXT I
380    REM Swap values in arrays
390    FOR I=1 TO S
400       FOR J=1 TO S
410          C(I,J) = H(I,J)
415       NEXT J
417    NEXT I
420 NEXT K
430 REM If you like, you can make a movie of all frames
440 REM to show foam evolving.
450 PRINT "To plot the foam, place a dot wherever the"
460 PRINT "C array has a value of 1 in it as you scan"
470 PRINT "I and J from 0 to S."
480 END
```

Computer Plays Black Hole Game with Itself (Chapter 13)

```
/* Computer plays Black Hole Game with itself */
#include <math.h>
#include <stdio.h>
int flag2, flag, bcount, numworm, i, j;
int bound,posx,posy,xs,ys,x[2000],y[2000];
float d1,r;
```

```
main()
{    srand(123456789); /* seed random nubmer generator */
     bcount=0; bound=25; /* size of playing board */
     for(numworm=1; numworm < 1000; numworm++) {
          /* randomly select first point of worm */
          x[bcount] = bound*(float)rand()/32767.;
          y[bcount] = bound*(float)rand()/32767.;
          bcount++ ;
          /* Determine 5 possible new points for worm */
          for(i=0;i<5;i++) {
               r = (float)rand()/32767.;
               posx=x[bcount-1];
               posy=y[bcount-1];
               if (r<=.25) posx = posx      +1;
               if ((r>.25) && (r<=.5))     posx = posx       -1;
               if ((r>.5)  && (r <=.75))  posy = posy       +1;
               if (r>.75) posy = posy         -1;
               /* make sure worm does not travel off board */
               if (posx <0) posx=0; if (posy <0) posy=0;
               if (posy >bound) posy=bound;
               if (posx >bound) posx=bound;
               x[bcount]=posx;  y[bcount]=posy;
               bcount++;
          }
          flag=0;  bcount--;
          /* scan to make sure worm does not intersect other
             worms */
          for (i=(bcount-5);i<= bcount; i++) {
               for (j=0;j< (bcount-5); j++) {
                    if ((abs(x[i]-x[j])<2) &&
                        (abs(y[i]-y[j])<2))
                         flag=1;
               }
          }
          /* scan to make sure worm does not intersect itself */
          for (i=(bcount-5);i<= bcount-1; i++) {
               for (j= i+1;j<= bcount; j++) {
                    if ((x[i]==x[j]) &&
                        (y[i]==y[j]))  flag=1;
               }
          }
          /* Call routine to make sure head is closer to black hole
             than tail */
          testdist();
          if (flag==0) {
               /* Print 6 points for plotting the worm body */
               for (i=(bcount-5); i<= bcount; i++)
                    printf("%d %d\n", x[i],y[i]);
               /* Print worm head at the following point: */
               if (flag2==0) printf("%d %d\n", x[i-1], y[i-1]);
```

```
                else printf("%d %d\n", x[i-6], y[i-6]);
            } /* if flag */
            /* if flag not right, worm intersected others or itself
               so reset counter */
        else bcount=bcount-5;
        } /*numworm */
}
/* Subroutine to determine if worm head is closer to black hole
   at center of board than tail */
testdist ()
{
    float cx, cy, d1, d2;
    flag2=1;
    /* position black hole in center: */
    cx = cy = bound/2.0;
    /* distance (squared) of tail to black hole */
    d1 = (x[bcount-5] - cx)*(x[bcount-5] - cx) +
         (y[bcount-5] - cy)*(y[bcount-5] - cy);
    /* distance (squared) of head to black hole */
    d2 = (x[bcount] - cx)*(x[bcount] - cx) +
         (y[bcount] - cy)*(y[bcount] - cy);
    /* flag2 is 0 if worm points to hole */
    if (d2 < d1) flag2 = 0;
}
```

Create Beautiful RLC Circuit Spirals (Chapter 14)

C Program Code

```
/* RLC Circuit Spirals */
#include <math.h>
#include <stdio.h>
float l, x, xnew, y;
int i,j;
main()
{
    l = 0.1; /* step size */
    srand(123456789); /* seed random number generator */
    for(j=0; j< 80; j++) {
        x = (6.*(float)rand()/32767.)-3;
        y = (6.*(float)rand()/32767.)-3;
        for (i=0; i<80; i++) {
            xnew = x + l*y;
            y = y + l*(-x - y);
            x = xnew;
            /* print out points along a spiral arm */
            printf("%f %f \n",x,y);
        }
```

```
      /* start a new spiral arm */
   }
}
```

BASIC Program Code

```
10 REM Compute Spiral Phase Portrait for RLC Circuit
20 L=0.1
30 FOR J=0 TO 80
40    X = (6.*RND)-3
50    Y = (6.*RND)-3
60    FOR I=0 TO 80
70       X1 = X + L*Y
80       Y  = Y + L*(-X - Y)
90       X = X1
100      REM PRINT POINTS FOR SPIRAL ARM
110      PRINT X, Y
120   NEXT I
130   REM START A NEW ARM
140 NEXT J
150 END
```

Mandelbrot Set (Chapter 14)

C Program Code

```
/* Compute Mandelbrot Set */
#include <stdio.h>
#include <math.h>

float xmin, xmax,ymin,ymax, newx, dx, dy, x, y, res, ru, ri;
int i,j,k, maxit, cutoff;
main()
{
    res = 100; /* image resolution */
    xmin=-2; xmax=2; ymin=-2; ymax=2; cutoff=4; maxit=100;
    dx = (xmax - xmin) / (float) res;
    dy = (ymax - ymin) / (float) res;
    for(i = 0, ru = xmin; (ru <= xmax) && (i < res); i++, ru += dx){
        for(j = 0, ri = ymin; (ri <= ymax) && (j < res ); j++, ri
            += dy){
            x=0; y=0;
            for(k=0; k<=maxit; k++){
                newx = x*x-y*y+ru; y =2*x*y+ri; x=newx;
                if ((x*x+y*y) > cutoff)  goto Break;
            }
            Break:
```

```
/* Give coordinates of black pixels (1) and white (0) */
if (k == (maxit+1)) printf("%d %d 1\n",i,j);
                    else printf("%d %d 0\n",i,j);
    }
  }
}
```

BASIC Program Code

```
.hr left right
10 REM Compute Mandelbrot Set
20 REM R is picture resolution
30 R=100
40 D1 = 4/R
50 D2 = 4/R
60 FOR R1 = -2 TO 2 STEP D1
70    FOR R2 = -2 TO 2 STEP D1
80         X=0
90         Y=0
95         F=1
100        FOR K=0 TO 100
110            X1 = X*X-Y*Y+R1
120            Y = 2*X*Y+R2
130            X=X1
140            IF X*X+Y*Y > 4 GOTO 400
150        NEXT K
160        F=0
170        REM Give coordinates of black pixels (1) and white (0)
400        IF F = 0 THEN PRINT I;J;"1"
410        IF F = 1 THEN PRINT I;J;"0"
440    NEXT R2
450 NEXT R1
460 END
```

BASIC Computer Program with Graphics Commands

Note: I personally enjoy programming using the language C, and have produced all the Julia sets in the color plate section using my own custom C programs. However, I know most hobbyists use BASIC. Therefore, I've included a BASIC program that should allow readers to compute and display Julia sets. Computer-literate readers should be able to further modify it for their own use. The program has been tested with GW-BASIC, Quick-BASIC, VisualBASIC, and PowerBASIC on the IBM PC platform and with QuickBASIC on the Macintosh. Users may wish to adjust the screen mode and size in lines 170–190 for the various platforms. The program produces a Julia dendrite in about 12 seconds using PowerBASIC 3.0 on a 486-DX2/66 machine. The values for CR and CI may be adjusted to produce different

Julia sets. Professor J. Clint Sprott, well known for his work in strange attractors, wrote the following code:

```
100 REM Julia set BASIC program for C. Pickover by J. C. Sprott
110 CR = 0            'Real part of C
120 CI = 1            'Imaginary part of C
130 X1 = -1.5         'Boundaries of plot
140 X2 = 1.5
150 Y1 = -1.5
160 Y2 = 1.5
170 SCREEN 12         'Assume VGA color graphics mode
180 W% = 640          'Screen width
190 H% = 480          'Screen height
200 KMAX% = 64        'Bailout condition
210 FOR I% = 0 TO W% - 1
220     FOR J% = 0 TO H% - 1
230         K% = 0
240         C% = KMAX%
250         X = X1 + I% * (X2 - X1) / W%
260         Y = Y2 - J% * (Y2 - Y1) / H%
270         XS = X * X
280         YS = Y * Y
290         WHILE K% < KMAX%
300             Y = Y + Y
310             Y = X * Y + CI
320             X = XS - YS + CR
330             XS = X * X
340             YS = Y * Y
350             K% = K% + 1
360             IF XS + YS > 4 THEN C% = K%: K% = KMAX%
370         WEND
380         PSET (I%, J%), C% MOD 16
390     NEXT J%
400 NEXT I%
410 END
```

Black Holes Evaporate (Chapter 15)

BASIC Program Code

```
10 REM Compute mass and temperature of a black hole through time
20 REM As a black hole ages, the temperature explodes
30 REM as the mass evaporates to zero.
40 REM M0 is initial black-hole mass in units of solar mass
50 REM M0=100
60 FOR T=1 TO 40 STEP 0.01
70    M = (M0*M0*M0-3*T)**.333333
80    IF M < .5 THEN PRINT "Danger! Black hole about to explode!"
```

```
80      PRINT "Time: ";T,"Mass: ";M,"Temperature: ";1./M
90  NEXT T
100 END
```

C Program Code

```c
/* Compute mass and temperature of a black hole through time
   As a black hole ages, the temperature explodes
   as the mass evaporates to zero.
   m0 is initial black-hole mass in units of solar mass */
#include <math.h>
#include <stdio.h>
double m, m0, t;
main()
{
        m0=100;
        printf("Time, mass, temperature\n");
        for (t=0; t <=40; t=t+.01) {
           m = pow(m0*m0*m0 - 3*t,.333333);
           if (m<0.5)
              printf("Danger.  Black hole about to explode!\n");
           printf("%f %f %f\n",t,m,1./m);
        }


}
```

Cross-Section of Traversable Wormhole (Chapter 16)

BASIC Program Code

```basic
10 REM Draw Cross-Section of Traversable Wormhole
20 B0 = 1
30   FOR R=B0 TO 10 STEP .2
40        Z = B0*LOG(R/B0  + SQR((R/B0)*(R/B0) -1))
50        REM  Print data for plotting upper right
60        PRINT R;Z
70   NEXT R
80   FOR R=B0 TO 10 STEP .2
90        Z = -B0*LOG(R/B0  + SQR((R/B0)*(R/B0) -1))
100       REM  Print data for plotting lower right
110       PRINT R;Z
120  NEXT R
130  FOR R=B0 TO 10 STEP .2
140       Z = -B0*LOG(R/B0  + SQR((R/B0)*(R/B0) -1))
150       REM  Print data for plotting lower left
160       PRINT -R;Z
170  NEXT R
180  FOR R=B0 TO 10 STEP .2
```

```
190        Z =  B0*LOG(R/B0 + SQR((R/B0)*(R/B0) -1))
200        REM  Print data for plotting upper left
210        PRINT -R;Z
220   NEXT R
230 END
```

C Program Code

```c
/* Draw Cross-Section of Traversable Wormhole */
#include <math.h>
#include <stdio.h>
main()
{
        float b0, r, z;
        b0 = 1;
        for (r= b0; r <=10; r=r+.2){
            z = b0*log(r/b0 + sqrt((r/b0)*(r/b0) -1));
            /* Print data for plotting upper right*/
            printf("%f %f\n",r,z);
        }
        for (r= b0; r <=10; r=r+.2){
            z = -b0*log(r/b0 + sqrt((r/b0)*(r/b0) -1));
            /* Print data for plotting lower right*/
            printf("%f %f\n",r,z);
        }
        for (r= b0; r <=10; r=r+.2){
            /* Print data for plotting lower left */
            z = -b0*log(r/b0 + sqrt((r/b0)*(r/b0) -1));
            printf("%f %f\n",-r,z);
        }
        for (r= b0; r <=10; r=r+.2){
            /* Print data for plotting upper left */
            z = b0*log(r/b0 + sqrt((r/b0)*(r/b0) -1));
            printf("%f %f\n",-r,z);
        }

}
```

Notes

Preface

1. "Anyone who falls into a black hole will plunge into a tiny central region of infinite density and zero volume . . ." Although this statement is true for classical mathematical models of black holes, many researchers believe that when space curvatures reach the value of an inverse Planck length squared, then classical solutions are no longer accurate. In this realm of "quantum gravity" quantum corrections are probably of the same order as the classical curvature. We can only *speculate* as to the exact behavior of bodies taking the cosmic plunge.

Chapter 1: How to Calculate a Black Hole's Mass

1. I should emphasize that the formula used to find the mass of the black hole works for any gravitating body, and the formula comes from the work of Newton, not Einstein. This equation does not only apply to black holes.

   In both chapters 1 and 2, I am envisioning a static "Schwarzschild" black hole; however, in reality this idealized black hole may be hard to find because there seems to be no mechanism for removing angular momentum from a rotating star as it collapses into a black hole. More realistic black holes are expected to be Kerr or Kerr-Newman (with exceptionally little charge) as described in later chapters.

2. Other chapters will discuss the inaccuracy of simple Newtonian solutions when gravitation fields are much more intense than in this example and when much higher speeds are traveled by the scolex. The formula presented in this chapter is an excellent approximation if we are not too close to the black hole.

3. When I describe white dwarves and neutron stars, I should emphasize that there are separate Chandrasekhar limits for the two, and that *both*

kinds of stars are supported by the "exclusion principle": white dwarves by electron pressure, neutron stars by neutron pressure. Also, there is not an infinite succession of stars composed of ever-smaller subatomic particles, because once the size of the star shrinks below the Schwarzschild limit, absolutely nothing can stand up to the gravitation.

Chapter 4: A Black Hole's Gravitational Lens

1. "I feel as if I were trapped in a dark pipe." In principle, because light rays circle the black hole one or more times, the image of the sky will be repeated an infinite number of times, but with rapidly diminishing brightness.

Chapter 5: A Black Hole's Gravitational Blueshift

1. "You realize it's a dream, and you wish there were some safe way, in real life, you could sail beyond the event horizon, pierce it, view its singularity . . ." In actuality, your dreams of visually seeing the singularity cannot be fulfilled for a Schwarzschild black hole. The singularity is never visible, even inside the black hole, because at every value of the radial coordinate inside the black hole, light can come to you only from the outside of the black hole (and only directly along the radial vector passing through you to the center of the hole).

Chapter 6: Gravitational Time Dilation

1. I use the expression "time slows," although it may reinforce your intuitive notion that there is one true time, and that events slow relative to this absolute. However, according to general relativity there is no such absolute, but rather only relative rates at which time flows, and, once one crosses the event horizon, the notion of relative rates of time flow fails as well. Of course, this is a mind-boggler, but that is the fun of black holes.

Chapter 7: Anatomical Dissection of Black Holes

1. "You have just cut through the outer event horizon and into the inner horizon." Technically speaking, the inner horizon is not an event horizon (it does not hide events from anyone) but rather a "Cauchy horizon" and a surface of infinite blueshift. This means that any energy (light) falling into this black hole is infinitely amplified at that surface, and thus

that for a realistic black hole the energy density is singular there. Because energy causes curvature, might the curvature (gravity) also be singular there? The answer is probably yes.

2. I describe "overcharging" a black hole, resulting in a naked singularity; however, you might not be able to add enough charge to produce a naked singularity for a real black hole. Once you achieve an extremal charge, where the charge equals the mass in Planck units, it is probably physically impossible to shoot another charge into the black hole.

3. I refer to the rotating black hole as being the shape of a lemon, which is prolate, whereas a rotating black hole is actually oblate. Unfortunately, I can't think of any common fruit that is oblate. All that comes to mind is summer squash.

4. "Outside of the static limit, I can remain static with respect to the black hole." This refers to being static with respect to stationary observers infinitely far away, not with respect to the black hole. By "stationary with respect to the black hole," one generally means with respect to zero angular momentum observers, which are in fact rotating with respect to a static observer.

5. "To gaze at the singularity would be a special treat indeed, as the only way to see one is to get so close that it would be impossible for you to counteract its massive, crushing gravitational pull." See notes for chapter 5.

6. On the topic of "no hair," Kip Thorne once told conference attendees that the new theory of black-hole evaporation left a particle theorist friend "a shaken man," because this meant black holes could eat baryons (a class of elementary particles, including neutrons and protons). The conservation of baryon number was an unquestioned absolute until then. Here is a scenario: Take a black hole of small size and moderate temperature, and add, for example, a '57 Chevy, baryon number 10^{25}. Next, wait for the black hole to evaporate for a while, and you will be left with *exactly* the same black hole because black holes "have no hair." However, the universe will be short 10^{25} baryons!

There are many other mind-boggling features of rotating black holes. For example, if you are orbiting close to one—even well outside the ergosphere—you can feel as though you are stationary and feel no centrifugal force even though your "inertial frame" rotates with respect to the stars. This is known as the "dragging of inertial frames," and Einstein believed it confirmed "Mach's principle." Specifically, Mach asked, "If the universe were empty save for one planet, how would that planet know whether it was rotating?" According to Mach, distant bodies determined a planet's local inertial frame, but according to general relativity, a black hole can alter such a determination.

Chapter 8: Embedding Diagrams for Warped Space-Time

1. If my use of the phrase "lift off the $z = 0$ plane" seems a bit confusing, consider an analogy of a rubber sheet with a black hole in the middle of it. The formula for the shape of the sheet seems complicated because constants are necessary to smoothly join two very simple shapes to one another.

Chapter 9: Gravitational Wave Recoil

1. A national scientific facility called the Laser Interferometer Gravitational-Wave Observatory, or LIGO, will be used to observe gravitational waves, most likely from "compact binary systems" of black holes.

Chapter 11: Gravitational Distension Near a Black Hole's Heart

1. "Finally your squashed corpse will merge with the quantum foam within the singularity." The hypothetical quantum foam will actually start to be encountered (that is, it will have structures that are important to the fate of a macroscopic body) in the regime where the curvature is proportional to the inverse Planck length squared, rather than within the singularity.

Chapter 14: Mathematical Black Holes

1. For each c, start with $z = 0$. Repeat $z \rightarrow z^2+c$ up to N times, exiting the computer loops if the magnitude of z gets large. If your program has finished the loop, the point is probably inside the Mandelbrot set. If you have exited, the point is outside the set and can be colored according to how many iterations were completed.

Chapter 15: Black Holes Evaporate

1. Mounted on the wall just outside Kip Thorne's office was once a large, ugly fire extinguisher, with a sign explaining that it was a class C extinguisher, not to be used on oil fires. Shortly after the *Scientific American* publication on exploding black holes, the sign was amended to include, "not to be used on exploding black holes."

2. "Because the ejected particles carry energy and this corresponds to temperature . . ." Technically speaking, many other gravitational phenomena are also responsible for quantum particle production (for example, cosmological models produce particles); however, only black holes and other gravitational objects with horizons produce a thermal spectrum and can therefore be thought of as having a temperature. This is often intuitively explained by the fact that black holes hide information. This hiding of the microscopic state produces entropy for the black hole. Anything that has an entropy must also have a temperature, by the laws of thermodynamics. However, the mathematical origin of black-hole entropy is one of the eminent outstanding problems of black-hole physics. To sum up: Energy emission does not necessarily entail having a temperature. The fact that black holes emit like "black bodies" is a very special feature of black holes.

Chapter 16: Wormholes, Cosmological Doughnuts, and Parallel Universes

1. "I could survive if the black hole were large enough." In reference to traversing an Einstein-Rosen bridge, many physicists believe that only spacelike trajectories cross the bridge, and thus, it would be very difficult to make it through regardless of how large the hole was. However, the idea that there can be low tidal forces for big black holes is correct.

2. "I could go through the ring without encountering infinitely curved space-time!" Of course, this is speculative because there is probably a "curvature singularity" at the inner horizon.

Postscript 1: Could We Be Living in a Black Hole?

1. This outside region is often called "asymptotically flat" in embedding diagrams for black holes.

2. The geometry of space in the vicinity of a collapsing body will be significantly curved; nevertheless, it is still convenient to attribute a value to the radius r, of a sphere in curved space, even though r is no longer necessarily equal to the distance of the surface of the sphere to its center.

3. As a quick aside, we know from previous chapters that black holes themselves eventually die through evaporation. Here are some of the evaporation times for the various black holes:

Stellar black holes: 10^{67}years
Galactic black holes: 10^{97}years
Supergalactic black holes: 10^{106}years

(In a strange sense, the black holes will have released their energy and matter into the external universe becoming analogous to white holes releasing their matter.) As many physicists today believe protons themselves are not eternal and disintegrate after about 10^{32} years, we can suppose all living things will have dissolved away long before black holes released their energy.

4. "... there is no requirement for the laws of physics to break down as far as you are concerned ..." However, is it not true that if you straddle the event horizon, your heart cannot pump blood to the other side of your body?

5. Some astrophysicists may be frustrated by the discussion, "Could we be living inside a black hole?" For additional insight, I include the views of Warren G. Anderson, a doctoral student conducting research into quantum aspects of black hole space-time in the Physics Department at the University of Alberta:

> Could we be living in a black hole? The answer probably depends on your definition of *black hole*. The technical definition is an asymptotically flat space-time with an event horizon. Asymptotic flatness means that you can define a place of finite size where the curvature (gravity) is concentrated, and that as you travel away from that place, the curvature (gravity) continually diminishes. To illustrate asymptotic flatness, consider a ball on an infinite rubber sheet. Near the ball there is a region where the rubber sheet is very curved, but as you travel away from the ball the sheet looks more and more flat. The event horizon is the boundary of a set of events (points in the space-time) from which no signal (not even light) can ever travel out infinitely far into the asymptotically flat region, even if we let it travel for infinite time.
>
> According to this technical definition, black holes do not exist in the strictest mathematical sense, since the universe does not, according to our observations, have a region outside which there is no gravitating object. On the largest scales we can observe in our universe, matter appears to be evenly distributed in all directions around us, and this matter will be causing gravity (curvature) everywhere; therefore, the universe does not seem to be asymptotically flat. Thus, according to the strictest definitions, the universe

does not even *contain* a black hole. This means that our definition is too strict. Now let us relax this definition and consider a black hole a piece of space-time. This piece is sufficiently flat when we are sufficiently far from some region which contains most of the gravity—we'll call this region the "sufficiently flat region." Let us further insist that this piece of space-time also has a set of points from which no signal can reach the sufficiently flat region, even if we wait for a sufficiently long time. We will call the boundary of this region of no escape the black-hole horizon. Such a weak definition would not be a good basis for mathematical theories, but it better describes what might really exist. Let me call an object that obeys the second definition a "realistic" black hole, and the one from the strict definition a "theoretical" black hole.

A realistic black hole is approximately described by the mathematical models we use for theoretical black holes. These models tell us that the interior of a black hole is a strange place. It is necessarily collapsing, and it is not in any sense homogeneous (the same everywhere) or even isotropic (the same in every direction). However, our universe is expanding and seemingly extremely homogeneous and isotropic. Therefore, our universe does not behave as expected if it were inside a black hole.

People argue that we do not really know what the inside of a black hole "looks" like because the theories we know break down there. This is true near the singularity where the gravity (curvature) is very strong. However, objects inside solar mass black holes would spend most of their time in regions where the space curvature is small and where our theories are expected to work very well. The percentage of time that objects spend in the small curvature regions becomes larger as the black hole becomes larger. Therefore, the theoretical description for the interior of a black hole with the mass of the observable universe should hold well almost everywhere inside. If one trusts the theory on the outside enough to accept black holes, why would one distrust it in most of the interior?

I've never heard active black-hole researchers claim we could be living inside a black hole. The valid theories do not agree with the experiment. People who think we are living in a black hole are probably using a black-hole definition that is different from the theoretical model studied by active researchers in black-hole physics.

Further Reading

General

Gardner, M. (1981). Seven books on black holes. In *Science: Good, Bad, and Bogus*. Buffalo, New York: Prometheus Books.

Gribbin, J. (1992). *Unveiling the Edge of Time*. New York: Crown.

Halpern, P. (1993). *Cosmic Wormholes*. New York: Plume.

Hawking, S. W. (1988). *A Brief History of Time*. New York: Bantam Books.

Kaufmann, W. J., III. (1979). *Black Holes and Warped Spacetime*. New York: Freeman.

Luminet, J.-P. (1992). *Black Holes*. New York: Cambridge University Press.

Misner, C. W., Thorne, K. S., and Wheeler, J. A. (1973). *Gravitation*. New York: Freeman. (Excellent, comprehensive background for more technical readers. A mega-textbook on Einstein's theory of relativity, among other things. Lots of equations.)

Morris, M. S., and Thorne, K. S. (1988). Wormholes in spacetime and their use for interstellar travel: A tool for teaching general relativity. *American Journal of Physics*, 56: 395.

Thorne, K. S. (1994). *Black Holes and Time Warps: Einstein's Outrageous Legacy*. New York: W. W. Norton. (Highly recommended reading.)

Wald, R. M. (1992). *Space, Time, and Gravity*, 2nd edition. Chicago and London: University of Chicago Press.

Wheeler, J. A. (1990). *A Journey into Gravity and Spacetime*. New York: Scientific American Library.

201

Preface

Balyuzi, H. M. (1980). *Baha'u'llah, The King of Glory.* Oxford: George Ronald Publishing. (This book gives a historical description of *Siyah-Chal.*)

Chapter 1: How to Calculate a Black Hole's Mass

Cowen, R. (1995). New evidence of galactic black hole. *Science News,* 147(3): 36.

Powell, S. (1994). Star gobbler. (A black hole is identified in the core of the galaxy M87.) *Scientific American,* August, 271(2): 12–13.

Chapter 4: A Black Hole's Gravitational Lens

Abramowicz, M. (1993). Black holes and the centrifugal force paradox. *Scientific American,* March, 268(3): 74.

Peterson, I. (1993). Wrapping carbon into superstrong tubes. *Science News,* April 3, 143: 214.

Stuckey, W. M. (1993). The Schwarzschild black hole as a gravitational mirror. *American Journal of Physics,* 61(5): 448.

Chapter 12: Quantum Foam

I have discussed recipes for artistic quantum-foam embedding diagrams in:

Pickover, C. (1992). *Mazes for the Mind: Computers and the Unexpected.* New York: St. Martin's Press. See also: Pickover, C. (1993). Lava lamps in the 21st century. *Visual Computer,* Dec. 10(3): 173–177.

For an excellent general background on probabilistic shapes created from cellular automata, see:

Vichniac, G. (1986). Cellular automata models of disorder and organization. In *Disordered Systems and Biological Organization.* Bienenstock., E., Soulie, F., and Weisbuch, G., eds. New York: Springer.

Chapter 16: Wormholes, Cosmological Doughnuts, and Parallel Universes

The following are highly technical articles with weird titles:

Benford, G. A., Book, D. L., and Newcomb, W. A. (1970). The tachynonic antitelephone. *Physical Review*, 2D: 263.

Clarke, C. J. S. (1990). Opening a can of wormholes. *Nature*, 348: 287.

Frolov, V. P., and Novikov, I. D. (1990). Physical effects in wormholes and time machines. *Physical Review*, 42D: 1057.

Morris, M. S., and Thorne, K. S. (1988). Wormholes in spacetime and their use for interstellar travel: A tool for teaching general relativity. *American Journal of Physics*, 56: 395.

Morris, M. S., and Thorne, K. S., and Yurtsever, U. (1988). Wormholes, time machines, and the weak energy conditions. *Physical Review Letters*, 61: 1446.

Redmont, I. (1990). Wormholes, time travel, and quantum gravity. *New Scientist*, April: 57.

Roman, T. A. (1994). The inflating wormhole: A *Mathematica* animation. *Computers in Physics*, 8(4): 480.

Visser, M. (1989). Traversable wormholes: Some simple examples. *Physical Review*, 39D: 3182.

Visser, M. (1989). Traversable wormholes from surgically modified Schwarzschild spacetimes. *Nuclear Physics*, B328: 203.

Visser, M. (1990). Wormholes, baby universe, and causality. *Physical Review*, 41D: 1116.

Index